DATE DUE

MAR 8 1991	SEP 2 6 1997	
SEP 27 1991	MAR 1 6 1998	
NOV 1 2 1991	NOV 0 7 1998	
FEB 6 1992	JAN 0 5 1999	
FEB 2 6 1992		
MAR 8 1993		
	ILL FAU	
JUN 0 7 1993	36261431	
MAR 2 1 1994	11/21/07	
MAY 3 1 1994		
MAY 2 6 1995		
NOV 2 8 1995		
JUN 0 5 1996		
JUN 1 5 1997		
SEP 0 8 1997		

DEMCO 38-297

THE 3-D OSCILLOSCOPE:
A Practical Manual and Guide

HOMER B. TILTON

Prentice-Hall, Inc., Englewood Cliffs, New Jersey 07632

1726464⁹

3-90

Prentice-Hall International, Inc., *London*
Prentice-Hall of Australia, Pty. Ltd., *Sydney*
Prentice-Hall Canada, Inc., *Toronto*
Prentice-Hall of India Private Ltd., *New Delhi*
Prentice-Hall of Japan, Inc., *Tokyo*
Prentice-Hall of Southeast Asia Pte. Ltd., *Singapore*
Editora Prentice-Hall do Brasil Ltda., *Rio de Janeiro*
Prentice-Hall Hispanoamericana, S.A., *Mexico*

© 1987 by

PRENTICE-HALL, INC.

Englewood Cliffs, N.J.

Printed in the United States of America

ABOUT
THE AUTHOR

The author's work in 3-D CRT displays began in 1949 with the independent conception, design, and construction of an oscilloscope to generate stereo images on a pair of type 3JP1 CRTs. The early success served as an impetus to continue this work, first reported in the literature in 1966.[1]

A monocular 3-D display developed by the author was marketed by Optical Electronics, Inc., beginning in 1965. This is

---

[1]Tilton, H. B., "3-D Display," *Instruments and Control Systems*, Aug. 1966, pp. 83–85.

the *scenoscope*—so-called because it produces scenographic (perspective) projections as do today's computer-generated 3-D displays. The scenoscope uses all-analog circuitry as opposed to the mostly digital circuitry used with the bulk of today's computer-generated displays. The scenoscope is noteworthy for that reason; for its early appearance on the scene; and because it implements a large number of depth cues—typically four or more depending on the model.

In one scenoscope model, the observer is tracked and the resulting position information is used to modify the CRT image so that he can "look around" displayed 3-D visual objects.

From 1968, the parallactiscope 3-D oscilloscope became the focus of the author's 3-D CRT work. Descriptions of the design, construction, and use of a practical laboratory parallactiscope form the subject matter of the bulk of this volume.

This author has also written a book *WAVEFORMS: A Modern Guide to Nonsinusoidal Waves and Nonlinear Processes* for Prentice-Hall published in 1986.

ACKNOWLEDGMENTS

The author wishes to thank Stephen Benton for the encouragement he provided during work on the parallactiscope, and Daniel Hudgins for research assistance in finding certain reference material and for physical assistance with the 14-inch parallactiscope at the 1985 IEEE-SID-AGED meeting.

Also acknowledged for stimulating conversations on 3-D and related subjects are James Butterfield (in memoriam), John Caulfield, Robert Collender, David Goodman, Bruce Lane, Lowell Noble, and especially Bill Greenwood.

DEDICATION

This book is dedicated to all those pioneers of the past, present, and future engaged in the pursuit of the third dimension.

WELCOME TO THE WORLD OF 3-D OSCILLOGRAPHY!

This book is a practical guide that takes you into another dimension of oscillography. It chronicles the path of researchers' efforts that led to the present state of oscillography. It then shows you how to adapt your own standard oscilloscope to produce three-dimensional *holoform* (hologram-like) images, and how to use 3-D oscillography in solving a wide range of problems, in Chapters 13 and 14. These "problems" come in basically two categories: displays that allow you to analyze the operation of a device or system, and displays that synthesize a particular desired space curve or surface for study or educational purposes.

Examples of the first category are:

- Three-dimensional "waveforms" that show the current and voltage characteristics of a diode or other device versus time;
- Surfaces which are characteristic of devices requiring three variables to characterize them—an example is the saddle-shaped surface characteristic of analog-multiplier operation;
- Medical uses such as the display of vectorcardiograms (VCGs).

Examples of the second category include classic surfaces of solid geometry such as:

- Helixes, including the double helix of DNA
- Cones of two nappes
- Spheres, spheroids, and ellipsoids
- Conoids—a surprising surface "discovered" by use of a 3-D oscilloscope
- Hyperbolic paraboloids, and more

However, the "problem" that may be most exciting to solve could well be one of those discussed in Chapter 16:

- All-electronic real-time holoform displays
- Holoform TV, movies
- Computer-generated holoform 3-D, and so on

As one anonymous reviewer eloquently put it, "Fame and fortune await the one who finds out how to apply this (holoform) principle to consumer electronics." The book is designed to set you on a course in hot pursuit of these goals.

The book pinpoints key factors leading to the development of practical 3-D displays on cathode-ray tube screens. It begins with the invention of the cathode-ray tube screens and ends with a list of challenges for you.

The central focus is the detailing of steps by which you can build your own 3-D oscilloscopes; starting with a stereoscopic oscilloscope (stereo oscilloscope) and culminating with a parallactic oscilloscope called "parallactiscope." The parallactiscope produces real-time holoform images having the dramatic depth cues of stereo and movement parallax; the latter allows you to "see around" displayed images simply by moving to one side as you would with a hologram or a real object.

The parallactiscope differs from moving-mirror displays in that it produces a controlled parallax by "reconstructing" light-ray directions as does a hologram. Holograms use wavefront interference to that end; parallactiscopes use a direction-sensitive "spatial" filter. Thus, parallactiscope images are holoform whereas moving-mirror images are not. The differences have a practical significance, which is explained in Chapters 2 and 5.

You will not be building an oscilloscope, nor will you be modifying one. You will simply be constructing subassemblies that will enable your oscilloscope to produce three-dimensional spatial images. Only hand tools are required, and no exotic parts or materials are needed. Photographs and diagrams guide you through the construction process.

"Multiscopic" photo sequences of 3-D displays are given (primarily in Chapters 13 and 14). These are static representations of the dynamic patterns you will see when you build and operate your own parallactiscope. The book tells you how those displays were generated, so you can produce them for "live" viewing on your own 3-D oscilloscope. To see stereo in the photos in this book you may need a stereoscope; but to experience hologram-like sensations with live parallactiscope displays, all you need are two eyes. Even if you close one eye, you can still "see around behind" the live parallactiscope images simply by moving left or right.

Applications for the parallactiscope certainly exist not only in the engineering fields, but also in the optical, biological, medical, and even the psychological sciences—as well as in pictorial art.

In summary, this book is concerned with the developing science and art of spatial imaging as it pertains to real-time CRT displays. The book is designed to set you on a course in the direction of accomplishment of one or more of the "blue-sky" projects described in the final chapter.

Set your sights high; for time and again, it has been found that the limits we encounter are limits that we, ourselves, have set!

CONTENTS

APPENDICES

PART I

THE HOW
AND WHY
OF 3-D
OSCILLOGRAPHY

CHAPTER 1

2-D:
JUST
AN
APPETIZER

1.1 BIRTH OF THE CRT: CROOKES' TUBE AND BRAUN'S BOTTLE

The story of the birth and the development of the CRT and its uses is a story of human achievement probably undreamed of in Jules Lissajous' time (1822–1880); yet it was the oscilloscope that made his name a household word—at least in the electronics laboratories around the world.

In 1857, Lissajous demonstrated the looping figures formed when a sinusoid is plotted against a second sinusoid whose

3

frequency is related to that of the first by a ratio such as 2/3, 3/4, 2/5, and so on (a rational number). Even earlier, in 1815, Nathaniel Bowditch had demonstrated similar looping figures.

In 1878, Sir William Crookes discovered cathode rays. Crookes showed that they can throw a visible shadow onto the end of an evacuated glass enclosure—a "Crookes' tube." He placed a Maltese cross inside of the tube so its shadow would be cast onto the end of the tube. The small amount of natural fluorescence in the glass caused a visible shadow of the Maltese cross to appear.

In 1897, Karl Ferdinand Braun added a phosphor screen to Crookes' tube to form the first true CRT (cathode-ray tube). (J. J. Thompson, who is generally credited with the discovery of the electron, found that same year that cathode rays are *beams* of free electrons, and *not* actually rays.) In 1902, Braun added a magnetic deflection apparatus. As a result of his work, Braun is generally credited with the invention of the CRT and the *cathode-ray oscillograph*, as the oscilloscope was first called. Electromechanical oscillographs were in existence at the time,[1] so one might suppose that it was natural to think of the oscilloscope as simply a new kind of oscillograph.

Both Crookes' tube and Braun's *Röhre* used a cold cathode. In 1883, Thomas Edison had discovered the phenomenon of thermionic emission (hot cathode emission of electrons), and in 1904 J. A. Fleming (later Sir Ambrose Fleming) invented the thermionic diode or *Fleming valve*.

In 1904, A. Wehnelt discovered an improved form of thermionic emission, one using oxides. He added a control grid[2] and oxide filamentary cathode to the CRT, thus producing the first CRT usable in a practical application. As the new science of electron optics developed, electrostatic deflection plates and other electrodes were added.

---

[1]W. Duddell invented an early electromechanical instrument—the bifilar string oscillograph—in 1893.

[2]As a result, the German name for CRT control grid is *Wehnelt* electrode.

✓1.2 THE COURSE OF CRT AND OSCILLOSCOPE DEVELOPMENT

With the advent of commercial radio broadcasting in 1920, the oscilloscope took on new importance and its development began to accelerate. Amplifiers and a synchronizable time-base generator were added. When radar was invented just prior to World War II, the triggered oscilloscope or *synchroscope* was developed. Virtually all of today's oscilloscopes that contain time-base generators (TY oscilloscopes) use triggered sweeps, but the name *synchroscope* is no longer used in this context.

Early oscilloscopes used CRTs having 3-inch diameter round screens. Later, 5-, 2-, and 1-inch CRTs became available. Nine-inch CRTs became available in the 1930s. In the 1940s, RCA (Radio Corporation of America) manufactured an oscilloscope with a 7-inch screen.

At the other end of the scale, in 1953 James Millen Mfg. Co. manufactured a 1-inch basic oscilloscope (one without sweep circuits and amplifiers). This was designed to mount on a panel in the space required by the then-popular 2-inch analog meter movement. A photo of one of these is shown in Fig. 1.1.

Although CRTs with larger screens soon became available, the 5-inch size became the industry standard for oscilloscopes. It

Figure 1.1 One-inch oscilloscope (90901) with its power supply (90902). James Millen Manufacturing Co., Inc.

remains so even today, although rectangular screens are now the rule and round ones the exception.

Early CRTs used the P1 (yellow-green) phosphor. The P1 screen is still available, but the chemical formulation and processing method have been improved over that first P1 phosphor. Even so, it is still referred to as "P1." P-numbers are assigned on the basis of electron-optical characteristics and not strictly on the basis of chemical formulation.

With the development of television, the P4 (white), P23 (sepia), and P22 (tricolor) phosphors were developed. Radar led to the development of still other phosphors. Today, the P31 is a popular phosphor for oscilloscopes. It is brighter and more bluish than the P1 phosphor and has a shorter persistence. The P31 is nearly ideal for use in the parallactiscope 3-D oscilloscope because of its short persistence and good visual output.

A tabulation of phosphor characteristics from P1 to P57 appears in Table 1.1.

In addition to the phosphors registered with EIA/JEDEC (EIA: Electronic Industries Association/JEDEC-USA: Joint Electron Device Engineering Council UK: Joint Electronic Defense Executive Committee), the various CRT manufacturers may have their own special phosphors.

Some of the first commercial manufacturers of cathode-ray tubes were Western Electric Company, Radio Corporation of America (RCA), and National Union.

By 1950, Du Mont and other manufacturers produced CRTs that contained multiple electron guns (four guns in one type), and others that were usable at frequencies up to 1 GHz.

In this age of solid-state electronics, when the electron tube has been almost completely replaced by semiconductors, it is interesting to note that the first electron tube (the CRT) is still going strong! Whereas solid-state panels have been devised to perform some of the CRT's functions, the CRT does not appear to be in any immediate danger of disappearing from the scene. Indeed, one of the properties of the CRT that is crucial to the operation of the parallactiscope 3-D oscilloscope has not yet been duplicated by solid-state panels. That property is its ability to produce images having very short persistence.

TABLE 1.1   STANDARD CRT PHOSPHORS

EIA/JEDEC Registration Number	Color		Persis- tence	Uses
	Fluor.	Phos.		
P1	YG	YG	M	oscilloscope
P2	BG	YG	ML	oscilloscope
P3 (obsolete)				
P4 (sulfide)	W	W	MS	TV
P4 (Sil-sul)	W	W	M	TV
P5	B	B	MS	photo
P6 (obsolete)				
P7 (1st layer)	B	B	MS	radar
(2nd layer)	YG	YG	L	
P8 (obsolete)				
P9 (obsolete)				
P10 (a scotophor dark trace)				
P11	B	B	MS	photo
P12	O	O	L	radar
P13 (obsolete)				
P14 (1st layer)		PB	MS	radar
(2nd layer)		YO	L	
P15 (UV comp)	UV	UV	VS	photo, FS scanner
(vis. comp)	G	G	S	
P16	V, UV	V, UV	VS	photo, FS scanner
P17 (1st layer)	YB	B	S	radar
(2nd layer)	BW	Y	L	
P18*	BW	BW	M	TV
P19	O	O	L	radar
P20	YG	YG	M	radar
P21	RO	RO	M	radar
P22	tricolor	(R, G, B)	MS	TV
P23	sepia	sepia	M	TV

*P18 is also called P4 *silicate type.*

TABLE 1.1 *Continued*

EIA/JEDEC Registration Number	Color		Persis- tence	Uses
	Fluor.	Phos.		
P24	G	G	VS	FS scanner
P25	O	O	M	radar
P26	O	O	VL	radar
P27	RO	RO	M	TV monitor
P28	YG	YG	L	radar
P29 (alternate strips of P2 and P25)				
P30 (registration withdrawn)				
P31	G	G	S	oscilloscope
P32	PB	YG	L	radar
P33	O	O	VL	radar
P34	BG	YG	VL	radar, oscil- loscope
P35	BW	BW	MS	photo
P36	YG	YG	VS	FS scanner
P37	B	B	VS	FS scanner
P38	O	O	VL	radar
P39	YG	YG	L	radar
P40	W	YG	M	radar
P41 (UV comp)	UV	UV	VS	photo, FS scanner
(vis. comp)	OY	OY	L	radar
P42	YG	YG	M	radar
P43†	YG	YG	M	radar
P44	YG	YG	M	radar
P45	W	W	MS	TV monitor
P46	YG	YG	VS	FS scanner
P47	PB	PB	VS	photo, FS scanner
P48	YG	YG	VS	FS scanner

†Narrow emission line.

TABLE 1.1 *Continued*

EIA/JEDEC Registration Number	Color		Persis- tence	Uses
	Fluor.	**Phos.**		
P49	bicolor	(R, G)	M	graphics
P50	bicolor	(R, YG)	MS	graphics
P51	bicolor	(R, YG)	(M, MS)	graphics
P52	PB	PB	MS	photo
P53	YG	YG	M	cockpit displays
P54	tricolor	(R, G, W)	MS	graphics
P55	B	B	MS	projection TV
P56	R	R	M	projection TV
P57	YG	YG	L	radar

KEY

Colors: Fluor. = fluorescence; Phos. = phosphorescence; B = blue; G = green; O = orange; P = purple; R = red; W = white; UV = ultraviolet, Y = yellow; BW = blue-white; and so on.

Persistence: L = long, M = medium, S = short; V = very

Uses: FS = flying spot; TV = television

Miscellaneous: Sil-sul = silicate-sulfide; vis. = visible; comp = component

1.3 TODAY'S OSCILLOSCOPES

Two basic kinds of laboratory oscilloscopes available today are: (1) *TY oscilloscopes*, designed to plot one or more signals against time, and (2) *XY oscilloscopes*, designed to plot one signal against another. TY oscilloscopes contain a time-base generator to supply sawtooth or ramp signals to the horizontal axis, whereas XY oscilloscopes do not. TY oscilloscopes are generally supplied with a front-panel switch that allows them to be used in an XY mode. XY oscilloscopes generally have an intensity-modulation (so-called Z-axis) input, and these are also called *XYZ oscilloscopes*. There is a larger variety of TY oscilloscopes available today than XY oscilloscopes.

It might seem that XY oscilloscopes, as such, would not be needed because TY oscilloscopes can be built with that option. That idea does not always work out in practice, because the specialization and optimization requirements of circuits to perform each of those two functions are frequently incompatible. An exception is the dual-trace oscilloscope, which is becoming increasingly popular. This provides the option of displaying two waveforms (dual TY mode), or of plotting two external signals against each other (XY mode).

All oscilloscopes designed to display signals above 10 to 100 kHz use electrostatic deflection, as opposed to electromagnetic deflection because of its inherently higher speed.

Both TY and XY oscilloscopes produce 2-dimensional images. Their usefulness has been such that 2-dimensional displays have seemed to be adequate even up to the present time. But now with "3-D" computer-generated displays capturing more and more imaginations, that situation is changing.

1.4 TRENDS IN OSCILLOSCOPES

One would be hard-pressed to purchase a new oscilloscope today that has more than one tube—the cathode-ray tube. Solid-state circuitry is used virtually without exception. This changeover from tubes to transistors and ICs (integrated circuits) began simply by performing the amplification, sweep, and power supply functions using transistors, Today, the trend is toward the use of more and more digital ICs.

The *digital oscilloscope* is designed for the display of analog waveforms, in spite of its name. It is "digital" in that scale factors, legends, cursors, and other markings are added electronically to the on-screen display.

Storage oscilloscopes for continuous display of transient signals are enjoying continuing popularity since their introduction in the mid-1950s.

There are oscilloscopes using traveling-wave CRTs for display of microwaves. There are oscilloscopes having multicolor displays. There are special purpose oscilloscopes, such as curve tracers.

The vectorscope is a special purpose oscilloscope that has a polar display for displaying the chrominance component of color television signals.

There are recording oscilloscopes that enable the operator to capture and store waveforms and transients so they can be displayed later.

There are digital analyzers[3] designed specifically for troubleshooting digital systems. Whether one considers these instruments to be oscilloscopes, or simply special instruments using CRTs is a matter of personal preference. (Are digital signals "oscillations"?)

In today's market, there are oscilloscopes that contain no tubes—not even a CRT. These are still expensive and have limited capabilities, whether in resolution, brightness, frequency response, or some other parameter.

Oscilloscopes generally have small screens, the reason being that, historically, electrostatic deflection was only practical in small-screen CRTs.[4] But there is a relatively new kind of electrostatically-deflected CRT available today: It has a large screen and the deflection sensitivity of a small-screen CRT. It uses *post-deflection magnification*, which means that its small deflection angles are magnified in the drift space so as to fill the entire faceplate of a, say, 21-inch screen. As a result, we may be seeing oscilloscopes with larger screens in the near future.

Large-screen XY oscilloscopes using magnetic deflection are available but these have a limited frequency response; they will not accommodate the high-frequency signals that small-screen units routinely work with.

Another possible large-screen oscilloscope of the future would use a raster scan. Waveforms to be displayed would be "captured" in digital storage and almost immediately played back onto the raster. The use of a raster-scan format would permit the use of larger magnetically-deflected CRTs, while digital storage would allow very fast phenomena to be captured and displayed.

[3]Not to be confused with digital oscilloscopes.

[4]Although 10-, 12-, 16-, and 19-inch electrostatically-deflected CRTs were available prior to 1965 from Sylvania Electric Products Inc. and others, the large deflection voltages necessary for full-screen deflection severely limited their usefulness.

1.5 NON-OSCILLOSCOPE USES FOR CRTS

One of the most important uses for CRTs is in television. In television systems, CRTs are used in receivers, monitors, camera view-finders, and flying-spot scanners. V. K. Zworykin contributed much to modern CRT design in adapting it to the reproduction of television images. For this he is generally credited with the invention of the *kinescope* or picture tube. Zworykin also invented the iconoscope camera tube.

Camera tubes (iconoscopes, image orthicons, vidicons, and so on) are close relatives of the CRT, as are scan-converter tubes and monoscopes. (The monoscope is a television image generator that has its own built-in graphic—usually a test pattern.)

Perhaps the largest and most complex relative of the CRT is the *scanning electron microscope* (SEM). The SEM is able to image things as small as molecules. At the other end of the scale, the *electron-ray tube* is probably the smallest and simplest relative. The electron-ray tube (compare "cathode-ray tube") was used as an aid in tuning radios in the 1935–1955 era. Its popular name was "magic-eye tube" because of the resemblance of its screen or *target* to an eye. The electron-ray tube is no longer in use. One reason is that its power-supply requirements are not compatible with those for transistors.

Among the many kinds of instruments that use CRTs today are automotive analyzers; medical monitors; military and air traffic control radars; cockpit displays for aircraft, spacecraft, ships, submarines, and automobiles; digital computers and digital workstations.

The use of CRTs to produce real-time 3-dimensional displays may seem to be a very recent development. Actually, proposals for such use can be traced back to the 1930s. In the next chapter, some of those proposals and developments are discussed.

1.6 LISSAJOUS: 3-D IN DISGUISE

In closing this chapter, we return to a consideration of the figures of Lissajous and Bowditch.

For certain frequency ratios, the Lissajous figure resembles a sinewave wrapped around a cylinder giving it a 3-dimensional appearance. This effect is especially pronounced if the frequency of one sinusoid is changed slightly so the pattern drifts, or changes slowly. The cylinder then appears to rotate slowly through a third dimension carrying with it a rigid, unchanging sinewave. Figure 1.2 shows two scope photos of such a "Bowditch figure" taken at slightly different times. When viewed stereoscopically, Fig. 1.2 shows this sinewave on a cylinder as if it were truly a 3-dimensional figure. Any of the viewing methods described in Appendix A can be used.

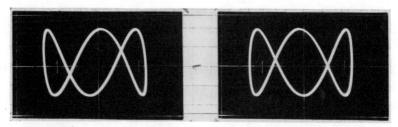

Figure 1.2   Two Lissajous figures that elicit stereo depth when viewed stereoscopically.

A way to obtain the stereoscopic effect with live oscilloscope displays is to view a slowly changing Lissajous figure through sunglasses having one lens removed while fixing your gaze on the center of the screen. By experimenting with the trace intensity and drift speed of the patterns, stereo depth can be seen. The stereo depth sensation so obtained is called the *Pulfrich stereophenomenon* or *effect*. The Pulfrich effect works because the eye that receives the dimmer image of the two "thinks" it receives the image later as well! That is, there is a *differential latent period* between stimulus and response times for the two eyes. This leads to a real image disparity on the two retinas, eliciting stereo depth.[5]

Upon first seeing slowly changing Lissajous figures on an oscilloscope, this writer became intrigued with the depth effect, and with consideration of how it could be enhanced and extended to other kinds of patterns. The remainder of this book is the result of an extended private research effort that began then, circa 1945.

---

[5]The Pulfrich stereophenomenon was discovered by C. Pulfrich in 1922. It has been studied extensively by Alfred Lit.

CHAPTER 2

3-D: THE MAIN COURSE

2.1 THE 3-D DISPLAY SPECTRUM

The various ways of generating 3-D displays, whether by using CRTs or some other imaging medium, can be categorized into positions along a continuum called the *3-D display spectrum*. At one end of that spectrum lie real objects and at the other end lie imagined objects. Intermediate points represent models having varying degrees of "real-ness." This spectrum appeared in a 1971 publication.[1] It is repeated here in Fig. 2.1.

Next to real objects are *physical models* (*see* Fig. 2.1). For example, a statue or other sculpture can give the same 3-D visual effect as the thing it models, although animation now becomes a formidable challenge.

Optical models are next to physical models. If a line or point of light can be made to come from an arbitrary location in space, then a given object can be simulated optically. One way to do this is to project a 2-D image onto a screen that is moving in a direction perpendicular to itself. This method is illustrated in Fig. 2.2. If the projected image changes in a controlled repetitive manner as the screen scans in a like repetitive manner, then a real 3-D image is thereby produced and a transparent 3-D object can be seen.

Alternatively, a stationary screen carrying an image can be viewed with a moving mirror or lens. This method is illustrated in Fig. 2.3. The image that is seen by looking in the mirror is virtual because it appears to lie behind the mirror even though it does not. The virtual depth excursion produced in this manner is twice the mirror excursion for a plane mirror, or even more for a magnifying mirror or one whose focal length varies repetitively as it scans—a *varifocal mirror*.

Next to optical models in the spectrum are *visual models*. These include stereoscopic, autostereoscopic, and holoform (hologram-like) images. Nearly everyone has seen 3-D movies. These are stereoscopic and generally require one to wear special eyeglasses or peer into a special viewing device. They do not permit the observer to "see around behind" visual objects by moving. That is, the "movement parallax cue" is missing. *Autostereoscopic* refers to stereo without special eyeglasses or other special viewing devices.

Holoform images, as the term is used here, are 3-D images produced using a stationary 2-D surface by generating a con-trolled parallax. This imparts stereo and movement parallax cues without special glasses or other observer constraints. The

[1]Tilton, H. B., "Real-time Direct-viewed CRT Displays Having Holo-graphic Properties." *Proceedings of the Technical Program, Electro-optical Systems Design Conference—1971 West,* Cahners Exposition Group, 1350 East Touhy Avenue, Des Plaines, IL 60017, 1971.

The 3-D Display Spectrum

The real thing	Physical models		Scanning optical models		Visual models		Neural models	The imagined thing
	Passive	Self-luminous	Real image	Virtual image	Stereoscopic	Holoform		
	Sculpture Wire, pin, and cardboard models Film stacks Gel-filled aquarium and moving stylus	Sparking or glowing wire intersections	Rotating EL panel Atomic resonance and crossed UV beams	Computer movies with varifocal mirror	Stereo movies (no observer movement please!)	Parallax-panoramagram Hologram Stereoptiplexer	Electrical stimulation of retinal receptors or cortical neurons	The imagined thing

—— Devices using cathode-ray tubes ——

	CRT with scanning screen	CRT with scanning plane mirror CRT with varifocal mirror	Stereo TV, radar display Stereo computer-generated display Stereo scenoscope	Parallactiscope
Features →	Automatically produces all 3-D visual cues except interposition	Virtual display space is larger than physically scanned region	All-electronic system using off-the-shelf hardware	Can display opaque objects No practical viewer restrictions
Limitations →	All displays are transparent No combination of ordinary TV cameras can provide input for 3-D TV image Display depth limited by size of device Sampling of input data necessary Requires high-output phosphor		Viewer restrictions; also, no movement parallax possible for multiple viewers	Cannot directly display low freq. information No vertical parallax Requires high-output phosphor

Figure 2.1 The 3-D display spectrum. Monocular 3-D (not shown) is a form of visual modeling. Adapted from the *Proceedings of the Technical Program: Electro-Optical Systems Design Conference—1971 West.* Used with permission. (Ref. 56.)

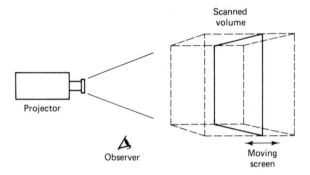

Figure 2.2   The scanning-screen idea.

hologram, properly illuminated, is an example of this kind of visual model. The parallax panoramagram is another. The hologram is discussed in Chapter 5. With the parallax panoramagram, specially prepared images are viewed through a thin lenticular sheet that is in contact with the image medium. The method is illustrated in Fig. 2.4.

Finally, between visual models and imagined objects are *neural models,* a possibility for the future suggested in 1963 by Allen B. Du Mont, wherein the brain is made to think that the eyes are seeing something. (Du Mont's original proposal was made in the context of a device to permit the blind to see.) Supposedly, an electronic means would be used to inject artificial signals into the neural channels. Ways of inducing hallucinations do not qualify, as the method must be exactly controllable and predictable, and must be non-damaging to the individual to have any practical value.

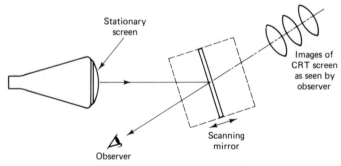

Figure 2.3   The scanning-mirror idea.

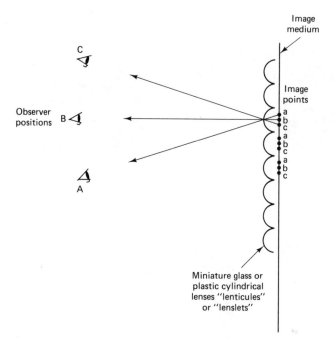

Figure 2.4 Illustration of the parallax panoramagram idea. Image points *a* are seen only at observer position *A*, and so on.

It should be noted that—insofar as visual perception is concerned—visual models can be just as real (in a 3-D sense) as physical models or even the real thing.

The remaining material in this chapter provides an overview of the pioneering work and some of the later developments in 3-D CRT displays.

2.2 HOW 3-D BEGAN

The term "3-D" is often taken to mean "stereo." Although it does not mean this, stereo is an important and powerful cue to depth perception. Consequently, this section is devoted mostly to stereoscopic and autostereoscopic methods of 3-D.

Prior to Sir Charles Wheatstone's invention of the stereoscope in 1838, the role of stereopsis in 3-D visual perception, or *space perception*, was not understood. Indeed, there appears to

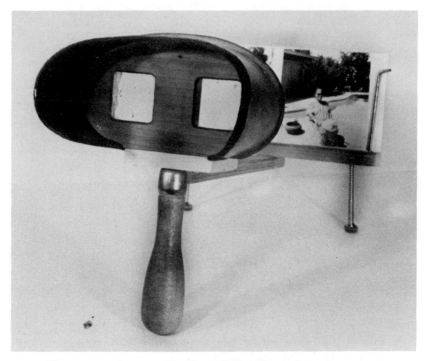

Figure 2.5   A reconstruction of an early parlor stereoscope.

have been general confusion on the point of how two eyes can collect a single 3-D image. Wheatstone showed, by his invention and application of the stereoscope, that this aspect of space perception is geometric and arises as a result of the subtle difference or *disparity* in the images collected by the two retinas. Photography was a relatively new art, and stereo photography became an exciting variation. Consequently, the Wheatstone device enjoyed popularity for a time as a parlor stereoscope. The author's reconstruction of one such device is shown in Fig. 2.5.

In 1891, nearly simultaneous suggestions were made by L. D. Duhauron and J. Anderton for anaglyphic[2] methods of

[2]**Anaglyphic** (adj.)—literally, *carved-up*. It is commonly used to refer to color separation of stereo images. Here we also use it to refer to polarization separation. Related methods of presenting stereo pairs are the use of *vectographs* (single pictures carrying a vector field of polarizations and intensities) and *time sharing*, wherein left and right images are presented alternately. In all of these methods, special viewing devices must be worn.

stereo. In these, the left and right images are superimposed in such a way that they can be separated and directed to the proper eyes by viewing in a particular manner. This would seem to be a definite improvement over Wheatstone's method, which required that the two images be printed side-by-side.

One anaglyphic method was demonstrated in a full-color movie at the 1939 New York World's Fair. Earlier in 1936, E. H. Land had found a way to mass produce the now widely used sheet polarizers. This made it possible, in the World's Fair demonstration, to project stereo movies and to view them using crossed polarizers to separate and direct the proper image to the proper eye of each member of the audience. This familiar technique is illustrated in Fig. 2.6.

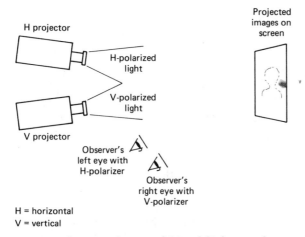

Figure 2.6   Showing how polarized light can be used to project and see stereo images.

The other anaglyphic method uses an image pair composed of complementary colors. The images are then directed to the proper eyes when the observers wear eyeglasses having like-colored filters for lenses. A. V. Bedford, in 1943, proposed using this scheme for stereo TV, and Sylvania Electric showed stereo TV images using this method in 1964. It has recently appeared on commercial TV. Since left- and right-eye images are in complementary colors (typically red and blue-green), this method is visually tiring because one color or the other tends to dominate in the observer's attention. This phenomenon is known as *color rivalry*. A clever proposal to eliminate it was made by E. T. Ferguson in 1968 using non-complementary colors constituted in such a way that they can still be separated by the use of optical filters. Spectra of possible colors to accomplish this end are shown in Fig. 2.7.

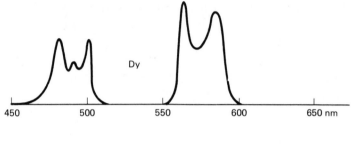

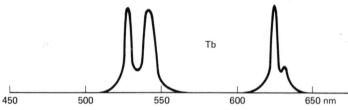

Figure 2.7 Complementary CRT phosphors. Dy is $Sr_3(PO_4)_2 - Dy$, Tb is $InBO_3 - Tb$ plus $Y_2O_3 - Eu$, rare earth phosphors.

Anaglyphic stereo movies in the commercial movie houses were popular in the early 1950s and again in the early 1980s.

V. K. Zworykin, in 1938, proposed a stereo television system in which scan lines are vertical, with alternate scan lines being keyed to left- and right-eye images. When viewed properly through a vertically-gridded mask or window, each eye would see only its intended image. Thus, stereo could be perceived in the displayed image. Observer position and image stability are critical with this method, but no special glasses need to be worn. Zworykin's scheme is illustrated in Fig. 2.8.

In addition to the proposals by Bedford and Zworykin, ways of displaying stereo TV images were proposed by C. W. Carnahan in 1942, J. J. Vanderhooft in 1957, A. Abramson in

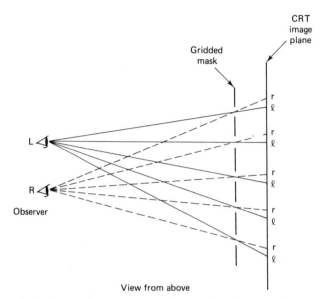

Figure 2.8 Zworykin's proposal for stereo TV display. Letters *l* and *r* represent left and right vertical scan lines. *L* and *R* are the left and right eyes of the observer.

1960, and C. W. Geer in 1965. Their proposals are diagrammed in Figs. 2.9 through 2.12.

Hughes Aircraft Co. (Vit) demonstrated a large CRT (21-inch round screen) stereo display showing 3-D radar images. In 1961, this writer was privileged to view live air traffic control situa-

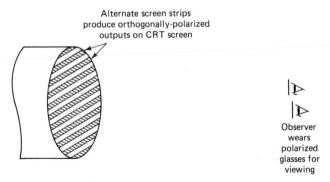

Figure 2.9 Carnahan's stereo CRT idea.

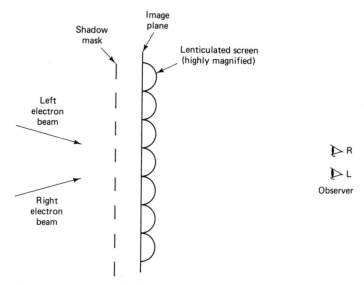

Figure 2.10 Vanderhooft's stereo TV idea. A stereo pair of images is produced using a screen structure resembling a parallax panoramagram.

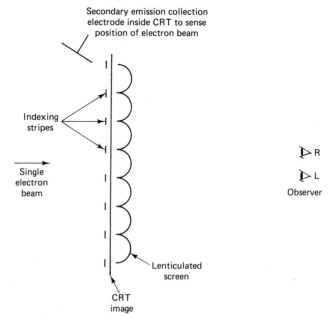

Figure 2.11 Abramson's stereo CRT idea. Color images are claimed as well as autostereoscopic ones.

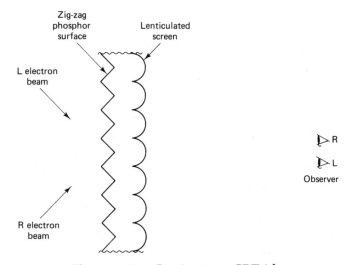

Figure 2.12   Geer's stereo CRT idea.

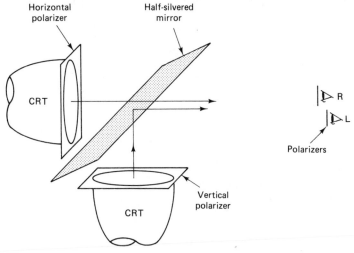

Figure 2.13   Hughes stereo idea.

tions on that display. There were two storage CRTs (*Tonotrons*[3])
with images combined anaglyphically using a mirror as diagrammed in Fig. 2.13. The multiple observers present were all
free to move. Viewing was accomplished by wearing eyeglasses

[3]Registered trademark of Hughes Aircraft Co.

having crossed polarizers for lenses. Unfortunately, the stereo-scopic depth effect was momentarily lost (to this observer) during times of movement, presumably due to the absence of the movement parallax cue.

An autostereoscopic method wherein the observer is tracked was reported by A. Schwartz in 1985.

Tektronix recently announced work on an anaglyphic CRT display that uses electronic switching of image polarization.

Two notable examples of holoform methods are the work of Roger de Montebello in 1977 with integral photography (an application of the parallax panoramagram), and of Robert Collender in 1965 with the stereoptiplexer and other photo-graphic systems. Collender's methods are unique in that they use a single vertical slit to produce a horizontal parallax. This idea is illustrated in Fig. 2.14. The slit is made to scan rapidly from side-to-side so that it effectively disappears, yet its parallactic effect remains. This moving slit will be referred to as a *Collender*

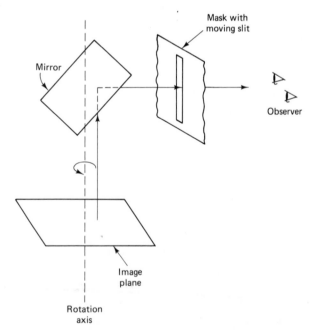

Figure 2.14 Collender's 3-D idea. The entire assembly (including the slit) rotates rapidly while the image on the image plane changes in synchronism.

spatial filter. It will be encountered again later in connection with the parallactiscope.

Both de Montebello's and Collender's works dealt with photographic imaging media. A recent news item tells of Ronald Kirk's work with another kind of spatial filter in conjunction with CRT images. He calls this an *optical tunnel array.*[4]

A CRT for painting real-time parallax panoramagrams of TV images appeared in a 1966 NASA Tech Brief.

There have been non-CRT methods proposed for producing physical models of 3-D objects. But in the CRT arena, the battle was joined early on by visual modeling on the one hand and optical modeling on the other. That battle is described in the next section.

2.3 THE EARLY BATTLE BETWEEN "SYNTHETIC" AND "REAL" 3-D

Circa 1960–70, optical modeling devices were known as "solid," "volumetric," "true," and "truly" 3-D, while visual modeling devices were known as "synthetic," "pseudo," "illusory," and "analog" 3-D. (Analog because they present *visual analogs* of 3-D spaceforms and scenes.) This nomenclature is probably the best evidence and indication we have today that optical modeling was considered to be the better way of the two to obtain 3-D images with CRTs at that time. But with the recent burgeoning of computer-generated 3-D displays—a form of visual modeling—that idea is rapidly fading.

In one corner of the ring, proposals for visual modeling ("synthetic" 3-D) were made as early as 1939, according to a 1948 report by E. Parker and P. R. Wallis. There was a proposal to use perspective projections of electronically-generated visual objects, that is, *computer-generated displays* in modern parlance. There was also a proposal to use stereo in 1942. Parker and Wallis pointed out that it is desirable to track the observer so the displayed images can be modified as required as the observer moves.

The use of oblique projections of 3- and 4-dimensional synthetically generated scenes was proposed by C. Berkley in 1948. Other proposals of this same nature were made by O. H.

[4]"Researching Holography." *International Television.* July 1985, p. 46. A discussion of Ronald Kirk's optical tunnel array.

Schmitt in 1947. Schmitt pointed out that displayed 3-D visual objects will be transparent unless unwanted points are specifically blanked out. This is the first mention found of the *hidden-line problem*, as it is called today.

The use of single-image projections with rotation capability is known as *monocular 3-D*, a seeming contradiction of terms. It is not really a contradiction; see Chapter 4 for further explanation.

In the other corner of the ring, CRT and other imaging devices for producing optical models ("real" 3-D), began appearing around 1960. Some of the early proposals are as follows.

In 1962, Richard D. Ketchpel proposed placing a moving screen inside the evacuated envelope of a CRT. Scanning electroluminescent (EL) and light-emitting diode (LED) panels have also been proposed. In another proposal, an aquarium-like tank would be filled with mercury vapor and caused to glow at arbitrary points and lines by crossed ultraviolet beams. These proposals are illustrated in Figs. 2.15 through 2.17.

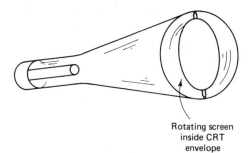

Rotating screen inside CRT envelope

Figure 2.15 Ketchpel's "solid" CRT idea.

A noteworthy scanning-mirror device is the varifocal-mirror display of A. C. Traub announced in 1967. This idea has been demonstrated both with computer-generated movies and with CRT imaging. It is diagrammed in Fig. 2.18. The mirror scan is cleverly produced by a mylar-mirrored membrane stretched across the mouth of a loudspeaker so that sinewave excitation of the speaker cone causes the mirror to alternately bulge and sag, thus varying the mirror focal length and producing the depth scan. The image is presented on film or a CRT screen and viewed via the varifocal mirror.

An all-electronic scanning-mirror device was recently announced by Tektronix. This uses several fixed mirrors (five in

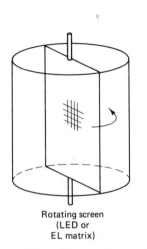

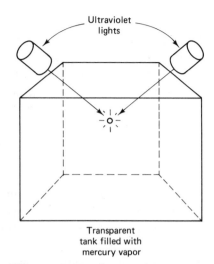

Ultraviolet
lights

Rotating screen
(LED or
EL matrix)

Transparent
tank filled with
mercury vapor

Figure 2.16 A variation of the scanning screen idea.

Figure 2.17 A mercury-vapor tank "solid" display idea.

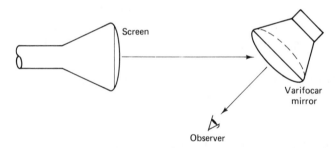

Screen

Varifocar
mirror

Observer

Figure 2.18   Traub's varifocal-mirror "solid" display idea.

one model) and electronically switches one mirror "on" at a time while the others are rendered transparent. Thus, a discrete depth scan can be produced.

2.4 ELATION, THEN DISILLUSIONMENT: THE "SOLID" DISPLAY HANGOVER

When first seen, the scanning screen and mirror proposals give one a feeling of serendipity—as if "Aha, this is it!" These "solid" displays seemed to be the answer to the quest for a practical real-time 3-D display.

But despite the considerable efforts expended by some of the corporate giants and others since about 1960, these display devices are still not available off the shelf. There are good reasons for this, one of which is the apparent requirement to have a mechanical moving component for a continuous depth scan. Another reason is that the amount of depth is apparently limited by the physical size of the display hardware. But perhaps the most devastating blow is that opaque visual objects cannot be produced with such devices for multiple observers; not even for a single observer without tracking him or her. This does not mean to minimize the on-going efforts with varifocal mirrors and electronically-switchable mirrors.

With the invention of the laser and the resulting rapid development of holography, it became abundantly clear that a stationary screen can carry an image that has all the visual attributes of an actual 3-dimensional object or scene. (The parallax panoramagram had "proven" that concept earlier, but high-quality parallax panoramagrams were not widely known.)

It is now generally conceded that holography is a desirable way to proceed in the quest for real-time 3-D displays. And, by their close association, parallax panoramagrams and other holoform displays now become real candidates for this coveted spot in the marketplace of the future.

2.5 CURRENT TRENDS: MONOCULAR 3-D

For some decades, efforts have been made to produce holography in real time. Computer-generated holograms have been produced, but these require substantial time to generate and arrange for viewing.

Perhaps because of the elusiveness of real-time holography, the old idea of monocular 3-D has recently been dusted off and carried to an amazing state of development. A brief description of the course of this development is set down here for the record.

Some early (circa 1960) monocular 3-D CRT displays began to be used as navigation aids as, for example, on submarines. These were called "contact analog displays." These were (and still are) pictorial imaging displays that permit the submarine pilot to navigate and maneuver underwater while watching his contact analog display much as an aircraft pilot does by simply looking out the windshield. Artificial grids are displayed that move and turn in the same way the submarine moves and turns.

Similar kinds of electronically-generated displays are projected onto the windshields of aircraft. These *head-up displays* might show synthetically generated runway images as a landing aid at times of poor visibility. Sometimes the displays use holographically-produced optics, but this does not make them holographic real-time 3-D displays as they are sometimes reported to be.

This idea of contact analog displays and head-up displays has been developed into highly sophisticated forms in flight trainers, science fiction movies, video games, and various other computer-generated display applications. Their increased sophistication must be credited, in large part, to the ready availability of large-capacity digital memories used with today's computers.

In these modern versions of monocular 3-D, images can be made to move in natural-appearing ways under software or sensor control. Although they are "only" 2-D projections, their realism is such that they are commonly referred to today as "3-D displays." Indeed, the term "solids modeling," as used today, has nothing to do with scanning screens and mirrors. Instead it refers to monocular 3-D visual models of opaque solid objects.

Whereas today's solid models are 2-D projections of opaque 3-D visual objects, *wire-frame models* are 2-D projections of only the outlines of transparent 3-D visual objects.

Lest the idea conveyed here be that projection is the only way to produce monocular 3-D, the use of sections is also important. One example of this kind of 3-D imaging is the medical CT scanner (formerly computer-aided tomography, or CAT scanner). One use for the CT scanner is in displaying cross-sections of the brains of living human subjects. One method of sensing for this case is the use of magnetic resonance (formerly nuclear magnetic resonance, or NMR).

The production of physical and optical models requires only that physical material or lights be placed at the proper positions in space to duplicate the form of the real thing. By contrast, if we would produce visual models, then we must know a great deal about the operation of visual perception. That is the subject of the next two chapters.

CHAPTER 3

HOW
AND WHY
YOU SEE REAL
OBJECTS IN 3-D

3.1 UNDERSTANDING THE VISUAL PROPERTIES OF PHYSICAL SPACE

It is convenient to speak of visual perception in terms of the properties of physical space even though perception relates to objects and scenes rather than to space *per se*. Since the laws of physics and geometry are the same for all physical objects, it is as if the "container" (space) were being acted on or *transformed* by those laws.

Geometric effects are important in visual perception. Other important effects relate to intensity, focus, and color shifts. All of these effects are *cues to space perception*.

The forms and shapes projected onto the two retinas of the binocular observer are related to the viewed scenes by the laws of geometry as governed by the optical properties of the eyes and the intervening medium. The light energy (photons) absorbed by the photosensitive receptors in the retinas causes the production of electrical pulse trains. The amplitudes of those neural impulses are independent of the stimulus intensity, but their frequency increases almost in direct proportion to the logarithm of the stimulus intensity, at least at normal (photopic) levels of illumination.

The neural impulses are conducted across the retina through nerve fibers to the point of egress at the blind spot or *optic disc* where each fiber combines with the other fibers to form the *optic nerve*. The optic disc lies to the nasal side of the *fovea*. The fovea is a small region at the center of the retina having the greatest visual acuity or resolution. When we look directly "at" something, we cause its image to fall upon the foveas of the two eyes.

3.2 THE "WIRING" OF THE EYE–BRAIN SYSTEM

The visual information collected by the eyes is sent via the optic nerves to the higher centers in the brain: The *lateral geniculate body* and the *visual cortex*.

In the normal human observer, the nerve impulses originating in the photoreceptors of the two retinas are sent along bundles of nerve fibers (contained in the optic nerves) by what may seem to be a roundabout way. The two optic nerves join at the *optic chiasma*, then separate again into two new "cables"— the *optic tracts*. The optic tracts contain the same nerve fibers that were in the optic nerves, but in a different grouping. Nerve fibers originating in the nasal (inner) halves of the two retinas cross at the optic chiasma; nerve fibers from the temporal (outer) halves of the retinas do not. The left optic tract terminates in the left hemisphere of the brain, and the right optic tract terminates in the right hemisphere. The arrangement is shown in Fig. 3.1.

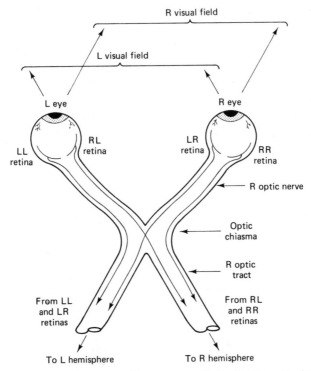

Figure 3.1 Schematic illustration of the "wiring" of the eyes.

If you look at Fig. 3.1, you will see that light from the right (or left) visual field registers on the left (or right) half of both retinas. The resulting neural impulses travel to the brain in such a way that image information from the right (or left) visual field is sent to the left (or right) hemisphere of the brain from both eyes.

3.3 COLLECTING 3-D INFORMATION ON 2-D RETINAS

The information from the two eyes, in the form of neural pulse trains, is combined or *fused* in the brain to form an information pattern having depth information as well as horizontal and vertical information. Each eye alone collects information of horizontal and vertical scene-point displacements. The difference in the horizontal information from the two eyes forms the basis for *stereo depth.*

The difference in the horizontal information collected by the two eyes occurs because of the *parallax* resulting from the difference in the horizontal position of the two eyes in the head. Since there is no difference in the vertical position of the two eyes, there is no difference in the vertical information collected by the two eyes. One important consequence of this is that *only horizontal parallax is necessary in a display to elicit stereo depth.*

3.4 PERCEPTION: HOW DOES IT OCCUR IN YOUR CENTERS OF COGNITION?

Although the precise mechanism of visual perception remains unknown, this much seems clear: In order for accurate perception to occur, the brain must faithfully decode the information sent to it by the eyes. That is, *it must produce an information pattern that duplicates the one contained in the original viewed scene.* Illusions are a result of an "unfaithful" or erroneous decoding operation.

We can find, by strictly analytical means, the information pattern sent to the brain. It is simply a result of the geometric and other transformations alluded to earlier. It follows that the brain (probably in the visual cortex) must perform the *inverse* of those transformations.

This highly simplified view of perception ignores the roles of learning and experience. Parametric inputs to perception undoubtedly also come from memory, expectation, and other brain functions. The question posed by the heading of this section is not a simple one to answer.

While we can feel pretty certain of what it is the brain must do in visual perception, how it does it remains a mystery.

In the next chapter, the geometric and other transformations that constitute the cues to depth perception are looked at in some detail. Those transformations are the things that must be understood in cold analytical detail if we are to devise realistic-appearing 3-D displays using the visual modeling approach.

CHAPTER 4

WHY YOU CAN SEE ARTIFICIALLY GENERATED IMAGES IN 3-D

4.1 DEPTH INFORMATION IS A SET OF "CUES"

The horizontal and vertical extensions of viewed objects are sensed directly. They produce corresponding horizontal and vertical image extensions on the retinas. By contrast, depth extensions are sensed indirectly through a system of depth cues.

The cues to depth perception can be categorized into two groups: *geometric cues* and *non-geometric cues*. Geometric cues are those which can be described in terms of geometric transformations. Among the geometric cues are *stereo, linear perspec-*

tive, observer–movement parallax, and gross scene rotation. Among the non-geometric cues are the inverse-square law of illuminance, hidden-line blanking (or interposition), aerial perspective, and depth of focus effects.

Stereo is a binocular depth cue; that is, it requires two coordinated eyes with overlapping fields of vision. All of the remaining depth cues just listed are monocular—requiring only one eye. Monocular depth cues form the basis for "monocular 3-D" introduced in Chapter 2.

Birds and fish navigate in 3-dimensional worlds. It is apparent that they must have excellent depth perception in order to do so successfully. It is a fact, however, that most birds and fish have their eyes placed on the sides of their heads with no overlap in the visual fields of the two eyes, resulting in no stereo! The fact that these creatures do navigate successfully is evidence of the power of monocular depth cues.

Motion that produces a gradual, continual rotation of the scene gives a powerful monocular cue to depth because it allows inspection of different views of the scene. It allows information about the shapes and relative positions of objects to be collected, resulting in 3-dimensional space perception without stereo.

Linear perspective is the name given to the reduction in size of retinal images as objects move away from the observer. Observer–movement parallax is the name given to the change in retinal image as the observer moves from side-to-side (an azimuthal change), up and down (an elevational change), or at any intermediate angle.

All eight of the above-listed cues are discussed in more detail in the sections immediately following this one.

4.2 HOW STEREO AND RELATED CUES WORK

In order to describe the geometric cues, we make use of a scene-centered rectangular Cartesian coordinate system tied to the observer. This coordinate system is shown in Fig. 4.1.

The observer's point of fixation[1] is at the origin of the coordinate system. The horizontal, vertical, and depth axes of the coordinate system correspond to the observer's horizontal, vertical, and sagittal axes, respectively.

[1]The point at which the two visual axes cross.

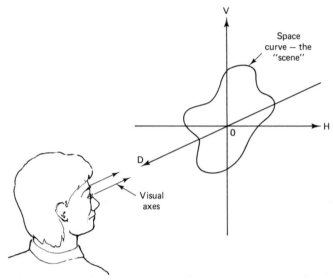

Figure 4.1 Scene-centered observer-based coordinate system.

Stereo and the other geometric depth cues can be described in terms of *linear transformations (LTs)*. In general, a linear transformation acts on an input vector (D,H,V) to produce an output vector (D', H',V'). In the present case, the input vector represents the scene information with D = depth, H = horizontal, and V = vertical. These ideas can be expressed conveniently and precisely in the language of *matrix mathematics*.[2] Specifically,

[2]The reader may recall that the matrix equation

$$\begin{bmatrix} x' \\ y' \\ z' \end{bmatrix} = K \begin{bmatrix} a & b & c \\ d & e & f \\ g & h & i \end{bmatrix} \begin{bmatrix} x \\ y \\ z \end{bmatrix}$$

has the same meaning as the simultaneous algebraic equations

$$x' = K(ax + by + cz)$$
$$y' = K(dx + ey + fz)$$
$$z' = K(gx + hy + iz)$$

The vector $\begin{bmatrix} x \\ y \\ z \end{bmatrix}$ is also written (x, y, z) in text.

The number K is called a *scalar*.

$$\begin{bmatrix} D' \\ H' \\ V' \end{bmatrix} = [\mathrm{LT}] \begin{bmatrix} D \\ H \\ V \end{bmatrix} \qquad (4.1)$$

where [LT] represents a 3×3 matrix. Specific LTs can be found by consideration of Fig. 4.1. They are as follows.

The LTs are, for the left (L) stereo image,

$$[\mathrm{LT}]_L = \begin{bmatrix} c & -s & 0 \\ s & c & 0 \\ 0 & 0 & 1 \end{bmatrix} \qquad (4.2a)$$

and for the right (R) stereo image

$$[\mathrm{LT}]_R = \begin{bmatrix} c & s & 0 \\ -s & c & 0 \\ 0 & 0 & 1 \end{bmatrix} \qquad (4.2b)$$

where s and c are the sine and cosine, respectively, of half the angle of convergence of the two visual axes. A description of how the stereo transformation is used to produce stereo displays on CRTs is given in the final section of this chapter.

Another LT is, for linear perspective (P),

$$[\mathrm{LT}]_P = \frac{1}{1 - z/A} \begin{bmatrix} 0 & 0 & 0 \\ 0 & 1 & 0 \\ 0 & 0 & 1 \end{bmatrix} \qquad (4.3)$$

where z is the depth coordinate of the scene point and A is the distance of the observer from the origin of coordinates (the point of fixation). This transformation is a scalar followed by an orthogonal projection. That is, it is just the factor $(1-z/A)^{-1}$ acting on the horizontal and vertical information. The depth information is not transformed because it is lost in the act of projection. This transformation is also called *perspective projection*.

For observer–movement parallax (OMP) the transformation is

$$[\mathrm{LT}]_{OMP} = \begin{bmatrix} \sec \phi_2 \sec \phi_3 & 0 & 0 \\ -\tan \phi_3 & 1 & 0 \\ -\sec \phi_2 \tan \phi_3 & 0 & 1 \end{bmatrix} \qquad (4.4)$$

where ϕ_2 is observer azimuth angle and ϕ_3 is observer elevation angle.

For gross scene rotation, the transformations are, for rotation about the vertical axis,

$$[LT]_V = \begin{bmatrix} \cos\theta & -\sin\theta & 0 \\ \sin\theta & \cos\theta & 0 \\ 0 & 0 & 1 \end{bmatrix} \tag{4.5a}$$

for rotation about the horizontal axis

$$[LT]_H = \begin{bmatrix} \cos\theta & 0 & -\sin\theta \\ 0 & 1 & 0 \\ \sin\theta & 0 & \cos\theta \end{bmatrix} \tag{4.5b}$$

and for rotation about the depth axis

$$[LT]_D = \begin{bmatrix} 1 & 0 & 0 \\ 0 & \cos\theta & -\sin\theta \\ 0 & \sin\theta & \cos\theta \end{bmatrix} \tag{4.5c}$$

where θ is each particular rotation angle.[3] Any two or all three of these rotation transformations can be cascaded (applied in sequence) to produce rotations about any two or all three coordinate axes. This corresponds to *multiplication* of the matrices. Chapter 15 discusses rotation in more detail.

Similarly, other transformations can be cascaded to produce combinations of geometric depth cues.

Perspective projection will be encountered again later in connection with the parallactiscope. In the parallactiscope, stereo and the horizontal (azimuthal) component of observer–movement parallax arise automatically as a result of the "parallactiscope principle" discussed in Chapter 6.

4.3 HOW OTHER DEPTH CUES WORK

The non-geometric depth cues are implemented by modulating CRT intensity, focus, and color (when applicable). The material in this section is written from the standpoint of the production of 3-D displays on a CRT since that is the context that most interests us.

---

[3]The three angles are called *Euler angles* after Leonhard Euler (1707–1783) who was the first to seriously consider them in analysis.

The inverse-square law of illuminance expresses the way perceived brightness of a point light source diminishes with distance. It can be implemented by applying a depth signal to the CRT control grid. It should be an inverse-square relationship, and should take into account the "gamma" (control grid and screen transfer characteristics) of the CRT. The polarity of the signal must be such that the intensity is less for those parts of the visual object that are farthest from the observer.

To produce hidden-line blanking (that is, to solve the "hidden-line problem"), an algorithm is required to determine when (at what points) the CRT is to be blanked. Such algorithms are part of the software packages for digital computer-generated 3-D displays. They are too complex to use with the parallactiscope and other 3-D oscilloscopes; however, a satisfactory approximation is available. This consists of using additional depth signal on the CRT control grid to produce a pronounced attenuation or blanking of those parts of a scene farthest from the observer.

Aerial perspective effects can be added by shifting the color of more distant parts of a scene towards the blue. This is applicable only to multicolor displays. Its purpose is to simulate the atmospheric effects encountered in looking at, say, a distant mountain range.

Depth of focus effects can be added by applying the depth signal to the focus electrode of the CRT. Thus the CRT focus control becomes a *focal plane control*. It moves the plane of sharpest focus in a direction toward or away from the observer. The amount of this cue can be adjusted by changing the amount of depth signal applied to the focus grid.

To further enhance the feeling of realism, highlights, shadows, and glint can be added by the proper use of intensity modulation.

Among the non-geometric depth cues, hidden-line blanking is routinely used with digital computer-generated displays—the others generally are not. All but color-shift effects are implemented in the scenoscope. None are implemented in the parallactiscope as described in this book.

In addition to the insertion of depth cues, the elimination of "flatness cues" enhances the realism of 3-D displays. By placing a window spaced a distance in front of the CRT screen, it becomes difficult to localize the actual position of the images, thus increasing the observer's conviction that these displays are real in a 3-D sense.

4.4 PRODUCING SYNTHETIC RETINAL IMAGES: HOW THEY RELATE TO REAL ONES

As an illustration of how to use the geometric transformations, a brief description of an early 3-D oscilloscope—the stereo oscilloscope—is given.

Stereo images are produced on a pair of CRTs or side-by-side on a single CRT. These become "synthetic retinal images" upon the act of viewing them. To produce realistic-appearing displays, these retinal images must correspond closely to the retinal images that would be obtained by looking at the actual object being simulated.

To produce the pair of CRT images for the stereo oscilloscope, the horizontal deflection input must be transformed before being applied to the horizontal deflection amplifiers. The vertical signal requires no change. This is attested to by the fact that multiplication of the matrices in the transformation produces the equation $V' = V$.[4] Thus the stereo transformations are effectively 2×2 matrices.

Starting with Eqs. (4.2), we drop the V component and factor c out of the LT, to obtain for the left (L) image

$$\begin{bmatrix} D' \\ H' \end{bmatrix}_L = c \begin{bmatrix} 1 & -t \\ t & 1 \end{bmatrix} \begin{bmatrix} D \\ H \end{bmatrix} \qquad (4.6a)$$

and for the right (R) image

$$\begin{bmatrix} D' \\ H' \end{bmatrix}_R = c \begin{bmatrix} 1 & t \\ -t & 1 \end{bmatrix} \begin{bmatrix} D \\ H \end{bmatrix} \qquad (4.6b)$$

where t is a small number (the tangent of the half angle of convergence) and c is approximately 1. We set $c = 1$ and

[4]This is a mathematical expression of the fact that vertical parallax is not required for stereo.

44   The How and Why of 3-D Oscillography   Part I

multiply the matrices in Eqs. (4.6) to obtain the two horizontal CRT signals, as follows:

$$H'_L = tD + H \qquad \textbf{(4.7a)}$$

$$H'_R = -tD + H \qquad \textbf{(4.7b)}$$

The D' signals aren't used. Equations (4.7) show that the left (right) image must have a small amount of the depth signal added to (subtracted from) the horizontal signal before application to the horizontal channels of the two CRTs.

A block diagram of the stereo oscilloscope using Eqs. (4.7) is shown in Fig. 4.2. The stereo oscilloscope accepts three waveforms as inputs and produces patterns containing stereo depth. For example, a helix can be displayed by inputting a pair of quadrature sinusoids and a sawtooth sweep.

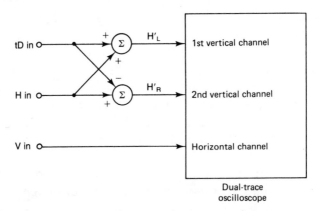

Figure 4.2 Block diagram of stereo oscilloscope. The oscilloscope's horizontal axis becomes the displayed vertical axis. This is accomplished by sitting the oscilloscope on end and viewing from the side using a stereo viewer.

In the stereo oscilloscope, implementation of Eqs. (4.7) produces images having a close visual effect to actual wire-frame models when properly viewed. The implementation of these transformations with electronic hardware can be accomplished by the use of operational amplifiers to add and subtract signals before going to the deflection amplifiers. This hardware can take the form of an adapter on the front end of a dual-trace oscillo-

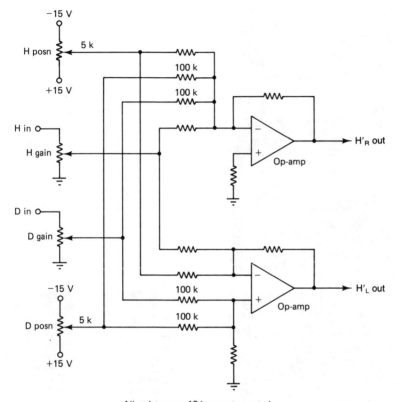

All resistors are 10 k except as noted

Figure 4.3 Schematic of stereo adapter for dual-trace oscilloscope.

scope. The schematic diagram of such an adapter is shown in Fig. 4.3.[5] Photos of some stereo pairs produced with this instrument are shown in Figs. 4.4 through 4.6.

The idea of synthetic retinal images applies as well to the parallactiscope and other 3-D displays that use the visual modeling approach. That is, we want to cause images to appear on the retinas that correspond closely to the retinal images obtained by looking at the object being simulated. We do this by implementing as many of the depth cues as required to produce the desired degree of realism.

[5]If your dual-trace oscilloscope has A−B and A+B channel functions, stereo pairs can be produced without the use of additional hardware.

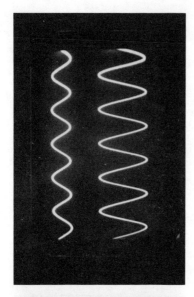

Figure 4.4 Stereo oscillogram (stereo oscilloscope display) of a sinewave on an inclined plane. View stereoscopically. Vertical format is necessitated by design of dual-trace oscilloscope.

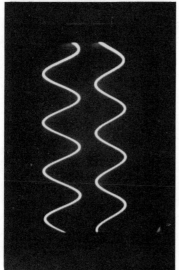

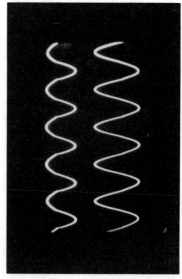

Figure 4.5 Stereo oscillogram of a helix.

Figure 4.6 Stereo oscillogram of a diode characteristic.

These synthetic retinal images are then displayed on a CRT screen or other image medium. They are transferred to the retinas simply by viewing them.

In the next chapter we look at holography and other holoform imaging techniques, and show how and why they automatically produce many of the geometric depth cues. Very soon after that, we enter into a detailed description of parallactiscope theory and construction.

CHAPTER 5

THE WHAT
AND WHY
OF HOLOGRAPHY
AND HOLOFORM
IMAGES

5.1 WHAT HOLOGRAPHY IS AND DOES

In 1948, Dennis Gabor described holography. In 1962, after the invention of the laser, E. N. Leith and J. Upatnieks perfected it. But what is holography, and why was laser light needed before it could be perfected?

To put it in simple terms, holography is a way of recording and reconstructing light ray directions. The astounding visual properties of the hologram are a direct result of that reconstruction. Holography is a way of photographically recording light diffraction patterns and subsequently "playing them back" to reconstruct the original ray structure. As a result, reproduced images have stereo, movement parallax, linear perspective, and other cues present in the original scene. That is, they are *3-dimensional.*

Lasers produce highly coherent light (light of a single wavelength) and so they produce the sharpest diffraction patterns. Laser light is useful for both recording and viewing, although the incoherent light from an ordinary incandescent lamp can be used to view certain kinds of holograms. Perhaps the best known of these is Stephen Benton's *rainbow hologram.*

The rainbow hologram uses the vertical direction on the film (by use of a special diffraction pattern) to analyze white light into its component wavelengths, thus producing a rainbow with horizontal bands. In this way, the incoherent light of the incandescent lamp is given a sufficient degree of coherence so it can reconstruct light rays in the horizontal direction to produce horizontal parallax, perspective, and stereo. There is no vertical parallax.

Holoform images are three dimensional because the light rays that reach the eyes come from the proper directions. The term "proper directions" is used for the directions that rays would assume if they were coming from a real object.

Holography is a way of reconstructing ray directions; but it is not the only way to reconstruct ray directions with the resulting "astounding" visual properties.

5.2 HOLOFORM IMAGES: THEIR PROPERTIES AND FORMATION

Holoform images were previously defined as "3-D images produced using a stationary 2-D surface by generating a controlled parallax." The conventional hologram and the rainbow hologram both fall into this category.

Holograms reconstruct light ray directions. They do this by using diffraction patterns that produce wave interference thereby reconstructing phase fronts. The eye is insensitive to phase-front information; so if ray directions can be reconstructed without

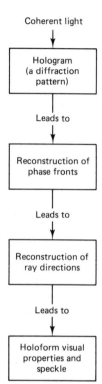

Figure 5.1 How holograms produce holoform visual properties.

first reconstructing phase fronts, the visual properties of holograms can be produced without relying on wave interference. Figure 5.1 diagrams the way in which a hologram produces holoform visual properties. In Fig. 5.1, it is shown that a properly illuminated hologram (a diffraction pattern) leads to reconstruction of phase fronts, which in turn leads to reconstruction of ray directions, finally leading to the production of holoform images plus *speckle*.

The presence of speckle in a hologram is a direct result of the use of coherent light, which is necessitated by the requirement to reconstruct phase fronts. Thus, if we can reconstruct ray directions without first reconstructing phase fronts, speckle can be eliminated from the resulting holoform images.

Parallax panoramagrams produce holoform images because they reconstruct ray directions. They do this without first reconstructing phase fronts. Therefore, they do not need coherent light and do not produce speckle. Instead, they rely on classical geometric (refraction) optics—on a microminiature scale. Their images are no less holoform than are those of holograms.

Holoform images are three dimensional because they reconstruct light-ray directions; but optical models (such as those produced by scanning screen and mirror devices) are three dimensional because they place points and lines of light at physical or virtual locations in space.

There is another property that sets holoform images apart from optical models. This property was mentioned earlier (in Section 2.4), but is important enough to be restated here in a slightly different form. It is that *the hidden-line problem can be solved without regard to observer position*. Holograms and parallax panoramagrams of solid objects are common forms; but optical models of solid objects have yet to be demonstrated. With such images, the points to be blanked depend on observer position. Thus it appears that it is impossible to display objects (using the optical-modeling approach) that appear solid for all observer positions.

Finally, holoform images that have only horizontal parallax (such as rainbow holograms) "follow you" as you move vertically. In that way they can be identified as being holoform.

5.3 PRODUCING HOLOFORM IMAGES WITHOUT RELYING ON WAVE INTERFERENCE

One method of producing holoform images without the use of wave interference is embodied in the parallax panoramagram. Other methods are also available.

Holograms and parallax panoramagrams use a static image medium—usually a rectangle of photographic film. But what if a dynamic image medium is used, such as a CRT screen?

With a CRT, we have the capability to write a rapid sequence of images in a short time. Thus, other methods of ray reconstruction are opened up to us. *We no longer need to produce all image elements simultaneously and continuously.* This eases considerably the required sophistication of the method needed to produce holoform images.

It now becomes possible to produce holoform images by use of parallax *at a point* or *at a line*. By scanning the point or line rapidly, all image elements can be built up in rapid succession. If a point is used, both horizontal and vertical parallax will exist in the manner of a conventional hologram. If a vertical line is used, only horizontal parallax will exist in the manner of a rainbow hologram.

That is the principle used in the parallactiscope. In the parallactiscope, parallax is produced along a vertical line. That line of parallax is then scanned rapidly from side to side so that it acts across the entire width of the image. Such images follow you as you move vertically, but you are able to look around them as you move horizontally.

5.4 THE PROMISE OF SYNTHETIC HOLOFORM IMAGES

A synthetic holoform image is one that is produced without using a real object or scene. Computer-generated holograms fall into this category.

Paintings, animated cartoons, and computer-generated images are familiar art forms. They permit expression of ideas without regard to reality. Synthetic holoform images take this progression one step further. They permit the expression of concrete pictures and abstract ideas in 3-D. The parallactiscope will introduce you to the world of synthetic holoform images. But the real promise of synthetic holoform images may lie in the direction of raster displays.

In a very real sense, oscilloscope images were forerunners of the complex CRT images that today's technology provides. A TY oscilloscope uses an arbitrary signal on the vertical axis, a sawtooth wave on the horizontal axis, and a pulse train to blank the retraces. A TV monitor uses sawtooth waves on both axes, a pulse train to blank the retraces, and a video signal to further intensity modulate the raster. A good case can be made for the claim that a TV monitor is simply a special purpose oscilloscope.

Thus the way to raster images follows readily from oscilloscope images, whether the oscilloscope is a conventional 2-D instrument or a 3-D parallactiscope.

Radar displays also evolved from oscilloscope displays. Indeed, any instrument that uses a CRT is connected—however tenuously—to Braun's original oscilloscope.

This book is about the 3-D oscilloscope. The promise of synthetic holoform images can be realized by starting with holoform oscilloscope images. When you have mastered this book, and built and used your own parallactiscope, you will be well versed in the production of synthetic holoform images. Then, when the world is ready for holoform TV, radar, and computer-generated displays, you will be able to show the way!

5.5 GENERATING HOLOFORM IMAGES IN REAL TIME

The capability of generating synthetic holoform images is certainly a fine thing. But the capability of generating them in real time is finer still.

The ingredients necessary to generate real-time holoform images are (1) a dynamic imaging medium such as a CRT, and (2) a way to produce a controlled parallax.

Consider this situation: You are using an analog voltmeter to measure a voltage. Because the meter needle is spaced in front of the meter card and scale, a parallax exists so that the reading you obtain depends on the viewing angle. The parallax in this case is undesirable, and some lengths are taken to remove it— such as the addition of a mirror at the meter card.

If we use a slit instead of a needle, and a CRT screen instead of a meter card, a parallax is produced again. This time the parallax is good because it can be harnessed to produce real-time holoform images. Consider the diagram in Fig. 5.2.

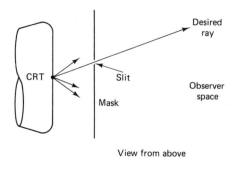

Figure 5.2 How to "reconstruct" light-ray directions with a slit.

Figure 5.2 shows the arrangement with a pinhole or slit spaced in front of a CRT. It is clear that the slit acts as a direction-sensitive *spatial filter*, which allows only particular light rays to escape into observer space while blocking all others. Thus, it "reconstructs" light-ray directions in a very real sense.

Since the CRT is a dynamic imaging medium, and since a parallax exists by virtue of the slit, all ingredients necessary to the production of real-time holoform images are present with this arrangement. How this is done is explained in the following chapter.

CHAPTER 6

THE PARALLACTISCOPE: A PRACTICAL 3-D OSCILLOSCOPE THAT YOU CAN BUILD

6.1 UNDERSTANDING PARALLACTISCOPE OPERATION

The Parallactiscope Principle

The *parallactiscope principle*, which governs the operation of the parallactiscope, can be understood by considering Fig. 6.1. From the figure, it is clear that each vertical line on the CRT screen is keyed to a unique viewing direction because of the presence of the slit. Thus, the CRT image seen by one eye of the

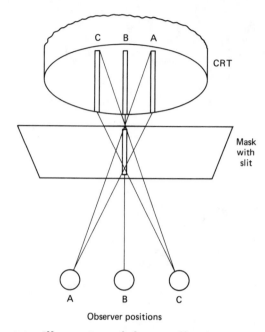

Figure 6.1 Illustration of the parallactiscope principle, showing how different observer positions are keyed to different parts of CRT screen.

binocular observer is not seen by the other eye. Indeed, an entire *continuum* of line images can be painted on the CRT screen, each of which is visible only from one direction of viewing.

Figure 6.1 illustrates the parallactiscope principle. How it is harnessed to produce images containing parallax is explained in the following paragraphs.

Obtaining Horizontal (H) Parallax

The situation existing in Fig. 6.1 is redrawn in Fig. 6.2 to emphasize the pertinent geometry. Referring to Fig. 6.2, the spot position Q at any instant corresponding to an arbitrary point P is easily found. A consideration of similar triangles shows that

$$\frac{h - s}{a} = \frac{x - s}{z'} \tag{6.1}$$

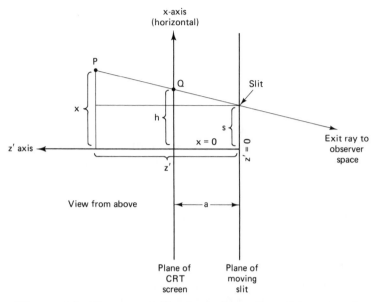

Figure 6.2 Geometry of horizontal parallax. Q is point of light on screen. P is point of simulated space scene.

This can be solved for h to obtain

$$h = \frac{x - s}{z'/a} + s \qquad (6.2)$$

In these equations, s is the instantaneous X-position of the slit, a is the distance between the CRT screen and the slit, and h is the instantaneous X-position of Q. The horizontal and depth coordinates of point P are x and z', respectively. Equation (6.2) will be referred to as the *H-parallax equation*.

For the moment, we let the vertical (v) signal equal the vertical (y) coordinate of point P.

Equation (6.2) is the relationship that must be electronically implemented in real time in order to paint the required image on the CRT screen from arbitrary x, y, and z' waveforms. Note that the horizontal signal (h) is a function of the slit position (s). This means that continuous information of the slit position must be fed to the circuitry that produces the h-signal.

A block diagram illustrating Eq. (6.2) is shown in Fig. 6.3.

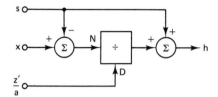

Figure 6.3 Block diagram of the H-parallax equation.

The H-parallax equation is good for the position of point P on either side of the slit (z' is negative in front of the slit). Notice that z' cannot go through the slit position because we cannot divide by zero.

Obtaining Vertical (V) Perspective

The H-parallax equation causes the automatic production of stereo, horizontal observer-movement parallax, and horizontal perspective. To produce vertical perspective, we must implement the vertical component of the perspective transformation, that is, Eq. (4.3). This is illustrated by the diagram in Fig. 6.4.

The V-perspective equation can be obtained by considering similar triangles. From Fig. 6.4 it is easy to see that

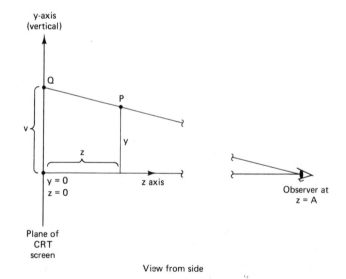

Figure 6.4 Geometry of vertical perspective. Q is point of light on screen. P is point of simulated space scene.

$$\frac{v}{A} = \frac{y}{A - z} \qquad \text{(6.3)}$$

This can be solved for v to obtain

$$v = \frac{y}{1 - z/A} \qquad \text{(6.4)}$$

In these equations, A is the nominal distance between the CRT screen and the observer, v is the instantaneous Y-position of Q, and y is the Y-coordinate of P.

In the case of the V-perspective equation, Eq. (6.4), it is convenient to take the depth coordinate as zero at the screen with the positive direction toward the observer. This is the Z-axis. In the case of the H-parallax equation, Eq. (6.2), it is convenient to take the depth coordinate as zero at the plane of the moving slit with the positive direction away from the observer. This is the Z'-axis.

The z and z' coordinates are related by

$$z = a - z' \qquad \text{(6.5)}$$

A block diagram illustrating Eq. (6.4) is shown in Fig. 6.5.

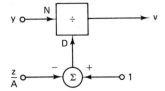

Figure 6.5   Block diagram of the V-perspective equation.

The V-perspective equation is good for the position of point P on either side of the screen (z is negative behind the screen).

6.2 THE PARALLACTISCOPE'S FIRST BABY STEPS

The first attempt to implement the parallactiscope principle was made in 1968 by modifying a Heathkit Model IO-21 3-inch oscilloscope. This oscilloscope (unmodified) is shown in Fig. 6.6.

In the modification, the 3RP1 CRT was replaced with a 3RP11 to provide the required shorter persistence. The CRT was moved back to allow a mask with a slit to be placed in front of the

Figure 6.6   Heathkit IO-21 oscilloscope used in the first parallactiscope.

screen but inside the cabinet. The mask/slit was placed on the end of a rod and caused to oscillate back and forth by two small speakers driven in push-pull at the mechanism's resonant frequency. (It was later found that one speaker was adequate for this task.) The arrangement is shown in Fig. 6.7.

The circuitry to implement H-parallax and V-perspective was placed in a case attached to the bottom of the oscilloscope's cabinet. This is shown in Fig. 6.8.

A next-generation parallactiscope was built from scratch around a 3JP11 CRT. A photo of this instrument is shown in Fig. 6.9. One significant improvement used in this unit was to replace the massive mask/slit arrangement with a small sliver of plastic.

Figure 6.7 Internal view of IO-21 modified to test the parallactiscope principle. Reprinted from the *Proceedings of the Technical Program: Electro-Optical Systems Design Conference—1971 West.* Used by permission.

Figure 6.8 The first parallactiscope disassembled. Re-
printed from the *Proceedings of the Technical Program:
Electro-Optical Systems Design Conference—1971 West.*
Used by permission.

A common method of forming an optical slit is to cut an
opening in an opaque mask. Another method is to place a sliver
of half-wave retarder between crossed linear polarizers. The
second method is now used in the parallactiscope because it
presents a much smaller mass to be moved (during scanning)
than the first.[1] If oriented properly, the retarder rotates the plane
of polarization of the light from the first polarizer by 1/2 piradian
(90°) so it is able to pass easily through the second polarizer.
Where there is no retarder, the light is blocked. Figure 6.10
shows this arrangement.

Limited success was obtained with these first two attempts.
Then we got smart. Instead of building a parallactiscope from

[1] It also occupies only one-third of the horizontal space required by the
first method while scanning.

Figure 6.9 Second parallactiscope built from scratch around a 3JP11 CRT.

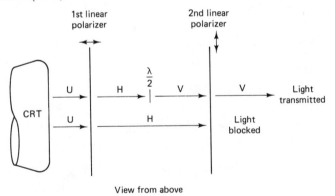

Key:

U = unpolarized light
H = horizontally polarized light
V = vertically polarized light
λ/2 = half-wave retarder

Figure 6.10 How a sliver of half-wave retarder can form an optical slit.

scratch or modifying an existing oscilloscope, it was decided to simply attach some kind of adapter to an existing XY oscilloscope.

6.3 THE 5-INCH PARALLACTISCOPE OF TODAY

Another Heathkit oscilloscope—a Model IO-4510—was purchased and assembled without any modifications. By this time, this manufacturer was using the P31 phosphor, which is almost ideally suited to parallactiscope use because of its short persistence and high intensity. Therefore, the CRT furnished with the oscilloscope was used.

An assembly was made for attachment to the top of the oscilloscope this time. It contained the slit scanner, its drive circuitry, and the circuitry for H-parallax and V-perspective. The slit mechanism was designed so the slit would hang down in front

Figure 6.11   The 1977 Mark 3 parallactiscope.

of the CRT. This instrument produced passable 3-D images, and it was demonstrated at a 1977 meeting of the SPIE. A photo appears in Fig. 6.11.

This instrument became the test bed for a "Mark 4" version. In the Mark 4, the slit scanner was kept in the subassembly atop the oscilloscope; but the circuitry for H-parallax and V-perspective was placed again in a separate subassembly. By this time, low-cost high-speed analog multiplier/dividers were readily available. The use of these greatly increased the dynamic range of the instrument, so that now it was easy to obtain virtual

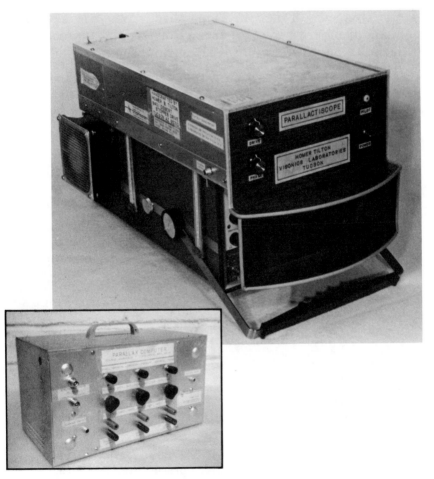

Figure 6.12   The Mark 4 parallactiscope.

depth excursions that exceed the horizontal and vertical
excursions. A photo of the Mark 4 device is shown in Fig.
6.12.

The Mark 4 instrument was simplified, cleaned up, and
became the instrument described in Part II—the parallactiscope.

6.4 WHAT YOU NEED TO KNOW TO DESIGN A LARGE-SCREEN PARALLACTISCOPE

A large screen (14-inch CRT) parallactiscope has also been
built and demonstrated. This uses a from-scratch approach, as a
suitable large-screen oscilloscope has not been available.

Figure 6.13   View of large-screen parallactiscope showing how the scanner pendulum passes around the CRT
from front to back.

The CRT is electrostatically deflected, a type 14E12P31M using post-deflection magnification (PDM) manufactured by Thomas Electronics, Inc. This parallactiscope is all electronic, except for the moving slit.

The moving slit consists of a lightweight aluminum and plastic torsion pendulum in the form of a large rectangle that straddles the CRT as shown in Fig. 6.13. By keeping the mass of this structure low, a resonant frequency of 4.7 Hz was obtained. This provides $2 \times 4.7 = 9.4$ scans per second since two scans occur for every cycle of oscillation. There is a resulting flicker, but it is not objectionable. Slit-scan peak-to-peak amplitude is 13-1/2 inches—more than adequate to encompass the full 11 inches of horizontal screen extent.

The remaining portions of this parallactiscope consist of deflection amplifiers and power supplies.

The circuitry required to solve the H-parallax and V-perspective equations is contained in a separate assembly. This is the same parallax processor that is used with the Mark 4 five-inch parallactiscope.

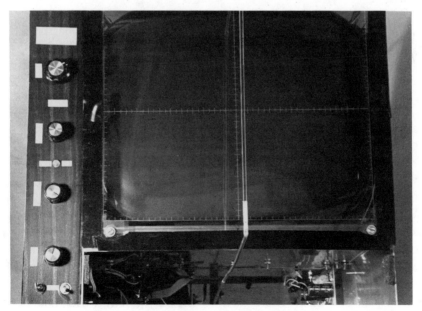

Figure 6.14   View of front showing slit.

Figure 6.15 Another view of the large-screen parallacti-scope.

Figure 6.14 is a close-up view of the front of the large-screen parallactiscope showing the slit, and the way one arm of the scanner pendulum passes under the CRT. Figure 6.15 shows another view of the large-screen parallactiscope outside its cabinet.

Figures 6.16 and 6.17 show the large-screen parallactiscope with its cabinet. The *panoply* or front hood contains the second linear polarizer. The first one is affixed near the CRT screen. Without the panoply and parallax processor, this instrument serves as a conventional XY oscilloscope. The slit mechanism can be left in place or removed during this application.

These descriptions leave some questions unanswered, such as, "How is the slit scanner driven at its resonant frequency?" This question and others will be answered when the details of parallactiscope construction are revealed next, in Part II.

Figure 6.16   The large-screen parallactiscope inside its cabinet.

Figure 6.17   Another view of the large-screen parallactiscope.

PART II

HOW TO BUILD YOUR OWN PARALLACTISCOPE

CHAPTER 7

TAKE
A LOOK AT
WHAT YOU WILL
BE BUILDING

7.1 A DECISION MUST BE MADE:
THE FROM-SCRATCH APPROACH VERSUS
THE ADAPTER APPROACH

Two approaches to parallactiscope construction are to build a complete instrument from scratch, or to build one or more assemblies to adapt a conventional XY oscilloscope. In the case of the 14-inch parallactiscope described in the previous chapter, there was no choice and the from-scratch approach had to be used.

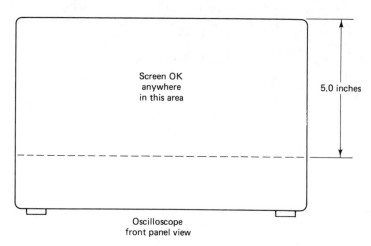

Figure 7.1   Screen limits

When a suitable XY oscilloscope is available,[1] the adapter approach results in simpler construction requirements. A "suitable" XY oscilloscope is one that uses a P31 phosphor in the CRT, has only a small phase shift (less than 3°) between horizontal and vertical channels, and has direct-coupled deflection channels.

One other requirement must be met by the "host" XY oscilloscope: Its screen must not fall outside the rectangle delineated in Fig. 7.1; or if it does, the design of the "parallax adapter" will have to be modified from that given in Chapter 9. If a handle or "back pack" is affixed to the top of the oscilloscope, it must be removed. An external graticule is preferable.

Other than that, virtually any oscilloscope can be used as the host instrument.

7.2 AN OVERVIEW OF ELECTRONIC AND MECHANICAL REQUIREMENTS

The three arbitrary waveforms that define the 3-D spaceform must be processed to provide the proper horizontal and vertical deflection signals for the CRT. This is one electronic requirement that must be met.

[1]Or one that can be used in an XY mode, such as a dual-trace TY model.

A slit must be provided that scans horizontally in front of the CRT. Various ways of implementing the moving slit have been considered. Among them were rotating discs and drums. These were rejected because they require multiple slits with their accompanying requirement to use high-precision fabrication techniques not generally available in the home workshop. It was decided to use a single slit and a sinusoidal scan because this is easy to implement, and has been shown to be effective.

Finally, a signal needs to be developed that accurately reflects the position of the slit from instant to instant. This is the s-signal required by the H-parallax equation.

7.3 HOW A MOVING SLIT IS USED TO PERFORM LIGHT-RAY RECONSTRUCTION

In the parallactiscope, the slit is carried on the front end of a torsion pendulum, which is oscillated to produce the slit scan.

It is desirable to keep the mechanical requirements at an absolute minimum to maximize reliability and to simplify construction. Therefore, a conventional motor is not used to drive the slit; instead, an ordinary loudspeaker is used to provide the motive force. By using a high-Q pendulum caused to oscillate at its resonant frequency, the speaker does not really have to "drive" the slit; it merely has to supply the small amount of energy that is lost in each cycle. This also results in insignificant vibration of the instrument.

To drive the slit pendulum at its resonant frequency, it is used as the frequency-determining element in a feedback oscillator. To do this, the s-position signal (s-signal) is used to provide the required feedback. The s-signal is obtained by means of a simple optical sensor described later. Since the slit motion is sinusoidal, the s-signal will be sinusoidal as well.

The width of the slit is not critical. A narrow slit (about 0.05-inch) provides high-depth resolution but low-image brightness. A wider slit (on the order of 0.10-inch) provides much better brightness and good depth resolution to a depth comparable to the horizontal image extent.

Spacing of the slit in front of the CRT was empirically determined for optimum results. It was found that with a 0.10-inch slit, a distance a equal to half the screen width is near optimum. Thus with the 5-inch parallactiscope we used $a \approx 2$ inches, and with the 14-inch parallactiscope we used $a \approx 5\frac{1}{2}$ inches. This distance is not critical.

With the slit this far from the screen, its scan amplitude (peak-to-peak distance) needs to be greater than the screen width. This is easily accomplished by putting a little more energy into the scan. Scan frequency is nearly independent of scan amplitude, so no problem is encountered from this direction.

7.4 OBTAINING AN ACCURATE MOVING SLIT WITH GARDEN-VARIETY HARDWARE

While half-wave retarder is best for the slit, you can find other materials that are nearly as good. Ordinary acetate serves well in this application. It should be thick enough to hold its shape but thin enough to possess a low mass. Material that is about 30 mils (thousandths of an inch) thick is good.

Experiment with different materials between crossed polarizers to see what gives the best transmission. Plastic from an ordinary two-liter soft-drink bottle works well; however, it is difficult to find a piece with surfaces that are as flat as might be desired. These materials have an axis because they are stretched during their manufacture. The material must be cut along the proper direction to provide maximum transmission.

Satisfactory sheets of linear polarizer can be obtained from auto supply houses. These are sold as sun visors; but not all sun visors are polarizing so test them against polarizing sunglasses or against each other before purchasing. Also try Edmund Scientific. Polaroid Corporation is the ultimate source for linear polarizers and half-wave retarders, but they have a $200 minimum-order requirement.

The pendulum that carries the slit material is made of small thin-wall aluminum tubing and music wire, available in many hardware stores and most hobby shops. This is glued together with epoxy and supported in a simple manner described later. A torsion pendulum requires a spring. This is fabricated from 1/16-inch music wire, and supported by the loudspeaker mounts. The pendulum is connected to the speaker by a yoke assembly, part of which is glued to the speaker and part of which is glued to the pendulum with silicone rubber. If you have ever built a model airplane, this part of the construction will be a snap.

The resulting pendulum oscillates in "simple harmonic motion," meaning it oscillates in an accurate sinusoidal manner. Thus, the slit is caused to move accurately from side to side in a sinusoidal manner.

7.5 THE SUBSYSTEMS: PARALLAX ADAPTER AND PARALLAX PROCESSOR

All of the required electronic and mechanical hardware is contained in two subsystems or subassemblies: a *parallax adapter* and a *parallax processor*.

The parallax adapter contains the moving slit and its drive electronics. This sits atop the host oscilloscope with the parallax-producing panoply (front hood) covering the CRT. Viewing is done by peering directly at the CRT screen through the front window of the panoply. This is the "second" linear polarizer shown in Fig. 6.10.

The parallax processor is a separate all electronic subassembly containing the circuitry for generating horizontal and vertical deflection signals from the three spaceform-defining waveforms and the s-signal. The parallax processor carries gain and translation controls for the three orthogonal input signals.

These two subsystems are described in some detail in the next chapter.

CHAPTER 8

HERE ARE
THE PLANS:
THE BLOCK
DIAGRAMS
AND SCHEMATICS

8.1 WHAT THE PARALLAX ADAPTER IS AND DOES: BLOCK DIAGRAM

The parallax adapter is shown in Fig. 8.1. Figure 8.2 shows it mated to the host oscilloscope—a Heathkit model IO-4510 in this case. Compare Fig. 8.2 with Fig. 6.12. Major components are identified in Fig. 8.3.

The parallax adapter is basically a 7-Hz electromechanical oscillator. Its purpose is to drive the scanning slit and to furnish an electrical output reflecting the instantaneous position of that slit.

Figure 8.1   The parallax adapter.

A block diagram of the parallax adapter is shown in Fig. 8.4. Referring to the block diagram, two photocells are eclipsed differentially by an oscillating shutter driven by a speaker (SPKR). The resulting signal is amplified by a differential amplifier (DIFF AMP). A balance adjustment potentiometer (BAL ADJ) provides for the signal out of the DIFF AMP to be set for zero d-c component (average value). The resulting signal is cleaned up and properly phased by a 7-Hz filter to form the s-signal.

The output of the DIFF AMP is phase shifted as well and used to drive the speaker via a power amplifier (PWR AMP). This phase shift is required because speaker current is out of phase with voice-coil movement.

The speaker is a 3- to 5-watt 4-inch square-frame unit. A lesser-power speaker might not furnish the required drive, and a greater-power speaker might contain a magnet so large as to affect the CRT electron beam. A square frame is required to obtain the required firm mounting and to properly accommodate the music-wire spring that you will fabricate and attach to it. A

Figure 8.2   Parallax adapter mated to host oscilloscope.

long-throw speaker is best used in this application, but is not mandatory.

The exciter lamp that supplies the required optical input to the photocells is energized by dc, as ac excitation would inject an unwanted signal into the system. The exciter lamp is operated at about half its rated voltage. This has two desirable effects: increasing lamp life by a factor of about ten thousand, and shifting the peak of radiation to more closely match the sensitivity of the two cadmium-sulfide photoconductive cells (photocells).

Cadmium-sulfide photocells are used because they are low cost, sensitive, and readily available—and because they have a time constant commensurate with the resonant frequency of the scanner pendulum. This prevents the pendulum from oscillating in modes other than the intended one.

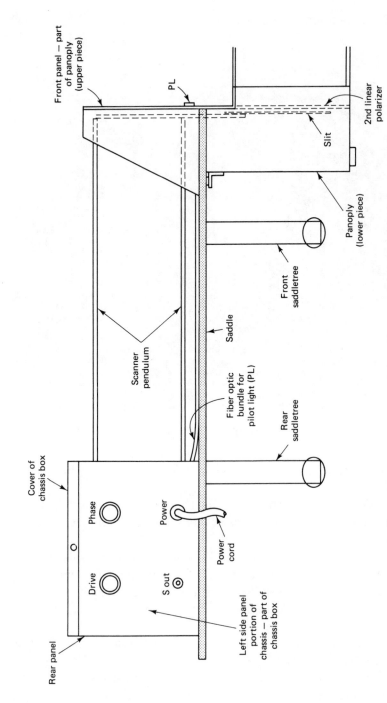

Figure 8.3 Major component identification—parallax adapter, left side view.

84

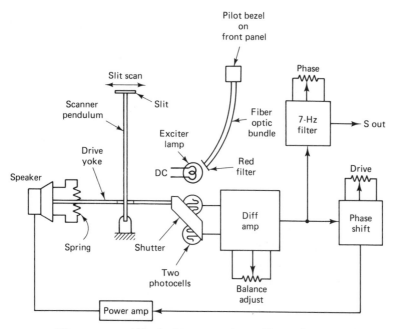

Figure 8.4   Block diagram of parallax adapter.

A fiber-optic bundle is used to pipe light from the exciter lamp to the front panel to serve as a pilot light and to indicate exciter-lamp operation. It is filtered red not only to contrast with the green CRT trace, but because that's the part of the spectrum in which most of its visible output lies.

8.2 SIMPLIFIED SCHEMATIC OF THE PARALLAX ADAPTER

A simplified schematic of the parallax adapter appears in Fig. 8.5. All active circuitry except the PWR AMP is based around type 747 dual operational amplifiers. U1A is a straightforward DIFF AMP that amplifies the signals provided by RΛ1 and RΛ2 photocells. The d-c level out of U1A is set to zero by R2, the BAL ADJ trimmer. The signal then follows two paths: One to provide the s-signal at J1, the other to drive the PWR AMP.

The s-signal to J1 is processed through an active bandpass filter, U1B, an *infinite-gain multiple-feedback* type such as

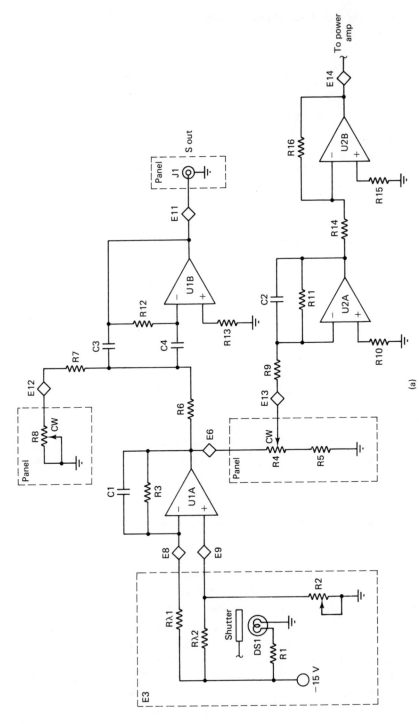

Figure 8.5 Schematic of parallax adapter. (A) Scanner amplifier. Components are on E4 except as noted.

86

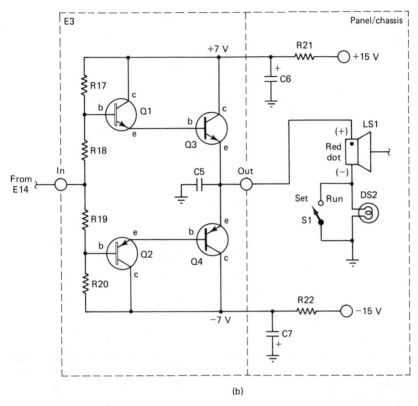

Figure 8.5 (B) Power amplifier (PWR AMP).

described by Johnson and Hilburn.[1] Components are chosen to fit the frequency of scanner operation, nominally 7 Hz. R8 is a phase control that allows the filter to be trimmed to give zero phase shift at the actual scanner frequency.

The other signal path, the one that drives the PWR AMP, processes the signal by phase shifting it in U2A and inverting it in U2B. The signal is then applied to the PWR AMP formed by Q1–Q4, and drives the speaker. The PWR AMP uses a Darlington complementary push-pull circuit, with direct coupling to the speaker. R4 sets the signal drive to the speaker to provide amplitude adjustment for the pendulum.

[1]Johnson, D. E. and J. L. Hilburn, *Rapid Practical Design of Active Filters* (New York: John Wiley & Sons, 1975), p. 153.

DS1, the exciter lamp, is energized from the $-15V$ supply via R1. DS2 is used as an indicator to show when R2 BAL ADJ is properly set.

8.3 WHAT THE PARALLAX PROCESSOR IS AND DOES: BLOCK DIAGRAM

The parallax processor is shown in Fig. 8.6. Its block diagram is shown in Fig. 8.7.

Referring to the block diagram, the parallax processor is based around two multiplier/dividers wired as dividers: one to implement the H-parallax equation and one to implement the V-perspective equation. The dividers are driven by three amplifiers that set gains and d-c levels. One of these processes the s-signal, one processes the horizontal (H) signal and one is used to set the amount of vertical (V) perspective and depth (D) centering. The depth gain potentiometer is tied in directly to the two dividers, as is the vertical (V) gain potentiometer.

Figure 8.6 The parallax processor.

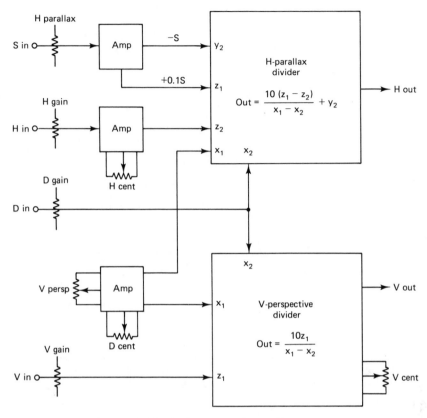

Figure 8.7   Block diagram of parallax processor.

8.4 SIMPLIFIED SCHEMATIC OF THE PARALLAX PROCESSOR

Figure 8.8 shows a simplified schematic of the parallax processor. Three type 741 operational amplifiers drive two Burr-Brown type 4213 multiplier/dividers.

The s-signal level is adjusted at the H PARALLAX trimmer to set the amount of H parallax to the proper level. This signal is then phase-inverted in U1 and applied to the Y_2 input of the H parallax divider, U2. The transfer characteristic of U2 is

$$\text{Ouput} = \frac{10(Z_1 - Z_2)}{X_1 - X_2} + Y_2 \qquad (8.1)$$

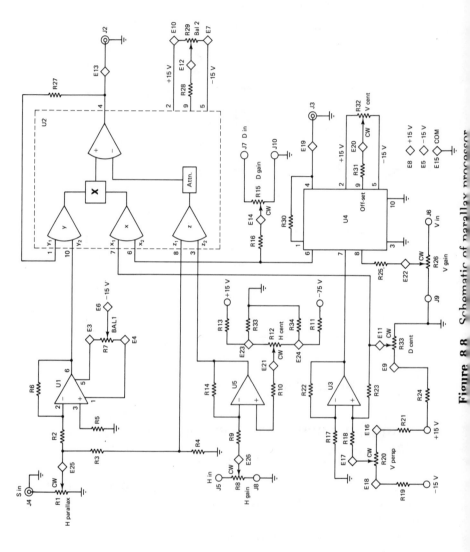

Figure 8.8. Schematic of parallax processor.

and $Y_2 = -S$ where S is proportional to the s-signal. One-tenth of S is also applied to the Z_1 input of U2.

Other inputs to U2 come from a network associated with U3 (the D-centering signal to X_1), the H-amplifier U5 (the horizontal signal to Z_2), and the depth input (depth signal to X_2). Thus, Eq. (6.2) is effectively implemented and the output of U2 is "right" to provide the horizontal input to the host oscilloscope.

Similarly, U4 provides a signal that is "right" for the vertical input to the host oscilloscope by combining vertical and depth signals properly. The transfer characteristic of U4 is

$$\text{Output} = \frac{10\ Z_1}{X_1 - X_2} \tag{8.2}$$

because Y_2 and Z_2 inputs are grounded.

Offset adjustments are provided at U2 (BAL2) and U4 pin 9. The one on U4 is a front panel control and serves as the vertical centering (V CENT) control. Other front panel controls are H CENT, D CENT, V GAIN, H GAIN, and D GAIN. Screwdriver adjustments are H PARALLAX, V PERSPECTIVE, BAL1, and BAL2.

CHAPTER 9

THE PARALLAX ADAPTER CONSTRUCTION DETAILS

9.1 HOW THE PARALLAX ADAPTER WORKS

The parallax adapter sits atop the host oscilloscope. Its panoply, containing the moving slit, covers the CRT screen. The front and rear saddletrees straddle the oscilloscope to position the parallax adapter on it properly. Refer to Fig. 8.3.

The scanner pendulum is caused to oscillate at a nominal frequency of 7 Hz through a peak-to-peak amplitude of about 7 inches at the slit end. The slit, carried at the end of the pendulum, consists of a narrow sliver of half-wave retarder supported by a lightweight structure of 1/16-inch aluminum tubing. The chassis box (refer to Fig. 8.3) contains all the electronics and all the oscillating mechanical components except that the pendulum extends from the front.

9.2 PARTS YOU WILL NEED TO BUILD ONE

Resistors, capacitors, and semiconductors you will need are listed in Table 9.1. Other parts are listed in Table 9.2. Materials are listed in Table 9.3. Knobs, wire, fastening hardware, solder, and so on are not listed. (All tables can be found at the end of this chapter.)

9.3 MECHANICAL CONSTRUCTION DETAILS

Saddle Subassembly

Construction begins by cutting a piece of 1/4-inch hardboard to the size and shape shown in Fig. 9.1 to form the saddle. Once

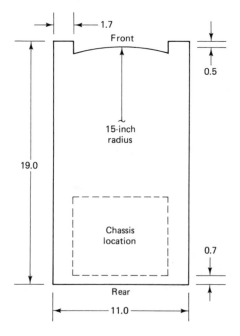

Dimensions are inches   **Figure 9.1** Saddle pattern.

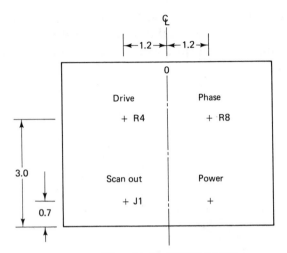

"Power" hole: drill for grommet

Dimensions are inches

(a)

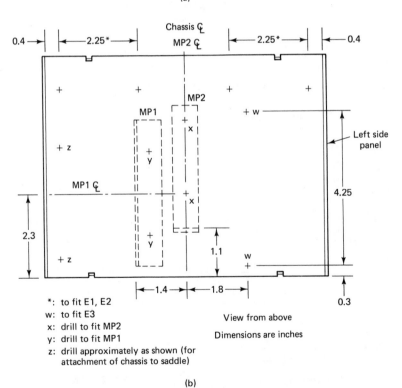

*: to fit E1, E2
w: to fit E3
x: drill to fit MP2
y: drill to fit MP1
z: drill approximately as shown (for
   attachment of chassis to saddle)

View from above

Dimensions are inches

(b)

Figure 9.2 Chassis drilling patterns. (A) Left side panel.
(B) Bottom plate.

that is cut, drill the chassis (lower part of chassis box) as shown in Fig. 9.2. Clamp this in place on the saddle as indicated in Fig. 9.1 and drill through the saddle using the same size drills you used for the chassis.

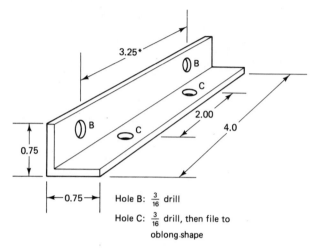

Hole B: $\frac{3}{16}$ drill

Hole C: $\frac{3}{16}$ drill, then file to
          oblong shape

*For $3\frac{1}{4}$ mounting centers on LS1

Dimensions are inches

Figure 9.3 Bracket MP1.

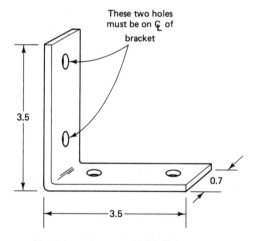

Material: steel, approx. 16 GA. Holes to
           accept #8 machine screws

Dimensions are inches   **Figure 9.4** Bracket MP2.

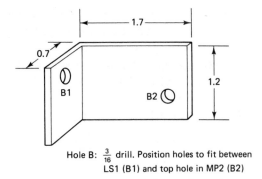

Hole B: $\frac{3}{16}$ drill. Position holes to fit between
LS1 (B1) and top hole in MP2 (B2)

Figure 9.5   Bracket MP3.   Dimensions are inches

Cut and drill brackets MP1, MP2, and MP3 as shown in Figs. 9.3, 9.4, and 9.5. Bracket MP1 is used to hold the speaker to the chassis and saddle. Bracket MP2 is used to hold the pendulum. Bracket MP3 is used to brace the upper end of bracket MP2 to the top of the speaker. These bracket dimensions are based on the

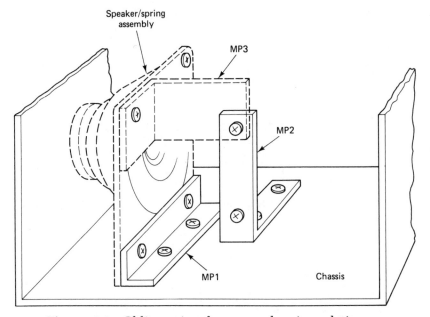

Figure 9.6   Oblique view from rear showing relative positions of Brackets MP1, MP2, MP3, and loudspeaker LS1.

speaker (LS1) listed in Table 9.2 having 3-1/4-inch mounting centers. If your speaker has different mounting centers, brackets MP1 and MP3 will have to be modified accordingly.

Bolt the chassis to the saddle at this time. Also, bolt brackets MP1 and MP2 to that subassembly as shown in Fig. 9.6.

Position this subassembly on top of your host oscilloscope so its center aligns with the center of the CRT. Make a sketch

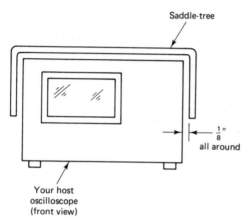

Figure 9.7 How to bend saddletrees.

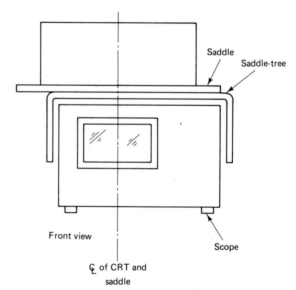

Figure 9.8 How to position saddle on saddletrees.

showing the relative positions of the saddle and the oscilloscope including dimensions.

Bend and attach the two saddletrees to the saddle as shown in Fig. 9.7 and in Fig. 9.8, using the sketch you made as a guide. Drill saddletrees and saddle approximately as shown in Fig. 9.9 for #8 machine screws, and fasten in place. Add cushions made of 1/8-inch felt or similar material as shown in Fig. 9.10. Position the subassembly atop the oscilloscope to be sure it fits snugly, but not too tightly.

Set this subassembly aside while you build the following subassemblies.

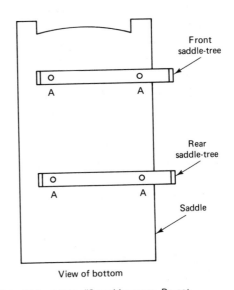

Figure 9.9 Hole locations for attaching saddletrees to saddle.

View of bottom

Holes "A": drill for #8 machine screw. Do not drill through chassis

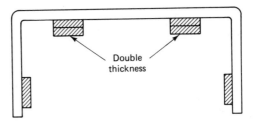

Figure 9.10 Felt locations on saddletrees.

 $\frac{1}{8}''$ felt

Scanner Pendulum

Lay out the plan for the scanner pendulum shown in Fig. 9.11 full size and tape it to a flat work surface. Cut and bend a piece of 3/32-inch music wire as shown. Slip on two 8-32 spade bolts, and tape it in place on your full-size layout.

Now cut and bend the thin-wall aluminum tubing as shown. Assemble the two horizontal beams with epoxy, and attach to the music wire with epoxy. Tape this subassembly to the pattern. Clean parts with alcohol before gluing, and use epoxy sparingly. Allow epoxy to set.

Add the two vertical slit supports as detailed in Fig. 9.12. It is especially important that this part of the structure be lightweight. The slit (half-wave retarder) will be added later, after the parallax adapter has been otherwise completed.

To secure the two spade bolts in place on the music wire, first bolt them to MP2. Turn the saddle assembly vertically, so the pendulum hangs from the two spade bolts. To center the music wire in the spade bolt holes, cut and slit two 1/8-inch lengths—3/32-inch inside diameter—of shrink tubing, thin-wall teflon tubing, or similar thin-wall tubing. Slip these in place

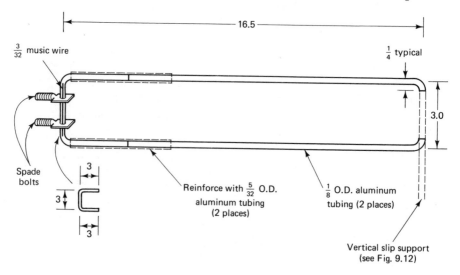

Dimensions are inches

Figure 9.11 Layout of scanner pendulum.

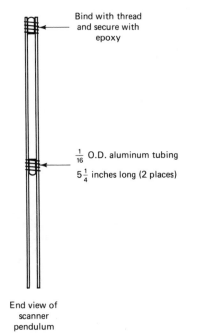

Bind with thread
and secure with
epoxy

$\frac{1}{16}$ O.D. aluminum tubing

$5\frac{1}{4}$ inches long (2 places)

End view of
scanner
pendulum

Spade bolts
and hex nuts

MP2

Fillet
here

Scanner
pendulum

$\frac{3}{8}$ inch

Down

Figure 9.12 Adding the two vertical slit supports to the scanner pendulum.

Figure 9.13 Position of pendulum prior to filleting of spade bolts.

around the music wire and center under each spade bolt. This is to ensure that there will be no metal-to-metal contact here.

Now, with the pendulum hanging and positioned as shown in Fig. 9.13, smooth enough silicone rubber (RTV) to form a fillet between each spade bolt and the music wire. Allow to set.

Speaker and Drive Yoke Subassembly

Cut and drill three pieces for the drive yoke from perfboard to the patterns shown in Fig. 9.14. Cut and form three lengths of 1/16-inch music wire as shown in Fig. 9.15. Slip yoke piece A onto music wire piece A as shown in Fig. 9.16.

Assemble the three pieces of music wire to form the pendulum spring by wrapping with #22 solid copper wire and soldering as shown in Fig. 9.17. Assemble this to the speaker with #10 hardware. Center and glue yoke piece A to the end of

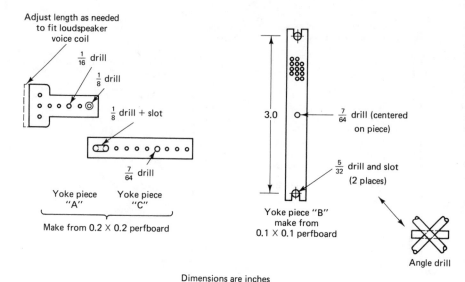

Figure 9.14   Yoke pieces A, B, and C.

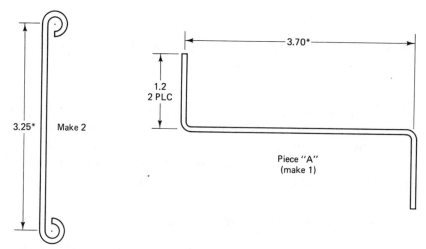

*For $3\frac{1}{4}$ mounting centers on LS1

Figure 9.15   Patterns for spring pieces. (Assemble per Fig. 9.17.)

the speaker voice coil with white glue. Trim or shim as necessary. (Shim with balsa wood.) Set this subassembly aside until the glue has set.

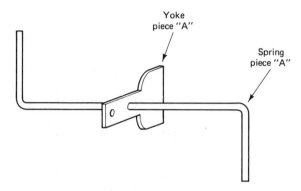

Figure 9.16   Assembling yoke piece to spring piece.

Assemble yoke piece B to the pendulum as shown in Fig. 9.18 and fasten with silicone rubber. Allow to set.

Mount speaker as shown in Fig. 9.19. Add MP3. Connectyoke piece C to pieces A and B as shown in Fig. 9.20. Apply silicone rubber as indicated in Fig. 9.20.

Panoply

The panoply is constructed from two pieces of 14-gauge aluminum; cut, bent and drilled as shown in Figs. 9.21 through 9.24.

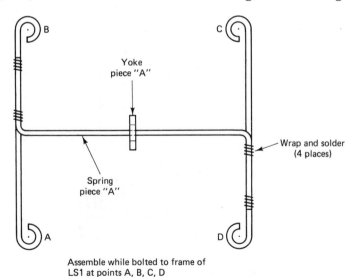

Figure 9.17   Assembling spring.

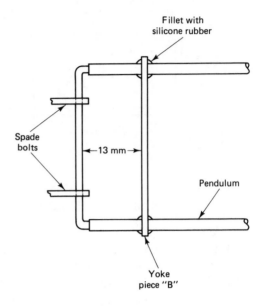

Figure 9.18 Attaching yoke piece to pendulum.

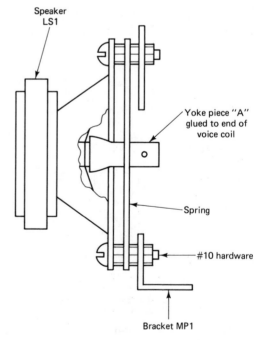

Figure 9.19 Mounting of speaker to bracket MP1.

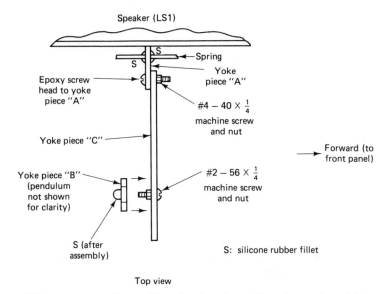

Figure 9.20   Assembly of yoke piece *C* to pieces *A* and *B*.

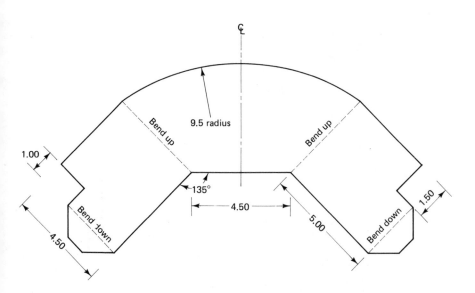

Bends are 90°, dimensions are inches

Figure 9.21   Pattern for bottom piece of panoply.

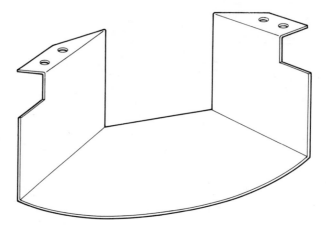

Figure 9.22 Appearance of bottom piece of panoply after bending. Drill 4 Holes for #8 machine screws approximately as shown.

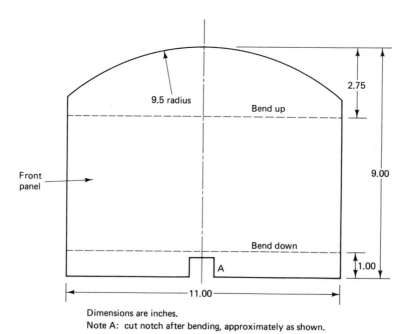

Dimensions are inches.
Note A: cut notch after bending, approximately as shown.

Figure 9.23 Pattern for top piece of panoply.

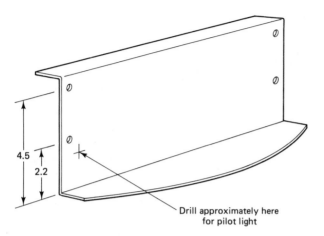

Dimensions are inches

Figure 9.24   Appearance of top piece of panoply after bending. Drill 4 holes for #6 machine screws as shown, .30 inch in from edge.

The bottom piece is fastened directly to the saddle as shown in Fig. 9.25. Fasten it now using #8 hardware. The top piece is fastened to the saddle using two right-angle brackets cut from 20-gauge aluminum and formed as shown in Fig. 9.26. Fasten these two brackets to the front panel (top portion of panoply) and

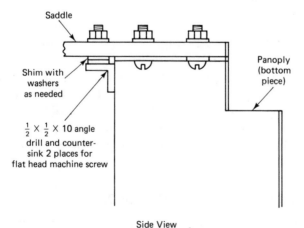

Side View

Figure 9.25   View showing how bottom piece of panoply bolts to saddle.

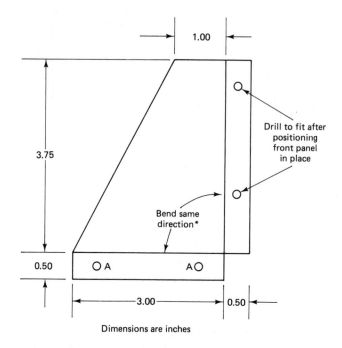

Dimensions are inches

Holes "A": drill approximately as shown
for #6 machine screws

*Bend to make 1 right-hand and 1 left-hand bracket

Figure 9.26 Pattern for brackets to hold top piece of panoply to saddle.

to the saddle as shown in Fig. 9.27 using #6 hardware. Rivets can be used to fasten brackets to front panel.

Add a piece of 1/2 × 1/2-inch aluminum angle, 10 inches long behind the lower portion of the panoply. This bolts to the saddle. It serves as a stop by butting against the front end of the oscilloscope. Add cushions at both ends so it will not mar the oscilloscope.

The linear polarizer will be added to the panoply and to the oscilloscope bezel later.

9.4 ELECTRICAL CONSTRUCTION DETAILS

Make printed circuit board E3 for the *scanner driver* now, using the pattern shown in Fig. 9.28. You can use the kit listed in Table 9.2. Helpful hints are given in Appendix B.

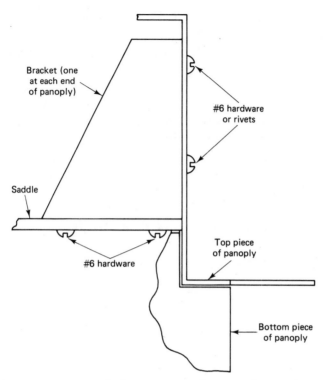

Figure 9.27 Detail showing attachment of top piece of panoply to saddle using angle brackets.

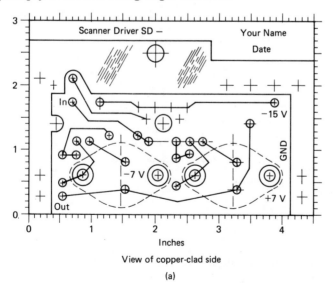

View of copper-clad side

(a)

Figure 9.28 E3 Fabrication. (A) Scanner driver layout.

After E3 is etched and drilled, mount and solder components according to Fig. 9.29.

Cut and drill E4 as shown in Fig. 9.30. Mount and solder components as shown in Fig. 9.31. This is the *scanner amplifier.* Mount two 6-32 spade bolts, and mount the scanner amplifier to the scanner driver as shown in Fig. 9.32. Connect wires between the two printed circuit boards as shown in Fig. 9.32. Mount this combination to the saddle and chassis as shown in Fig. 9.33. Mount J1, and mount PHASE and DRIVE controls (R8 and R4) to the left side panel. Mount S1 and DS2 to the right side panel as shown in Fig. 9.34.

Wire and solder C6 and R21 to terminal strip E1 as shown in Fig. 9.35, and C7 and R22 to E2 as shown in Fig. 9.36. Mount

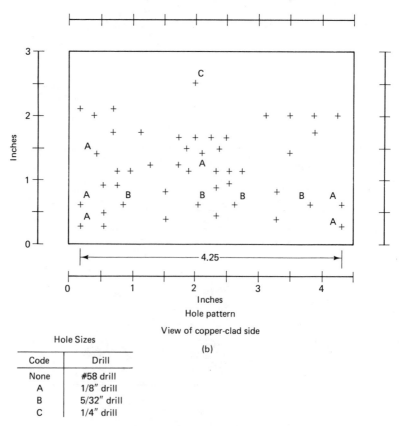

Hole pattern

View of copper-clad side

(b)

Hole Sizes

Code	Drill
None	#58 drill
A	1/8" drill
B	5/32" drill
C	1/4" drill

Figure 9.28 (B) Drill pattern for E3.

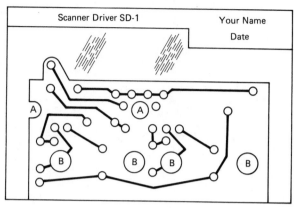

PC pattern

View of copper side (holes not shown)

Pad Sizes

Code	Pad dia.
None	3/32″
A	1/4″
B	5/16″

(c)

Figure 9.28   (C) Etch pattern for E3.

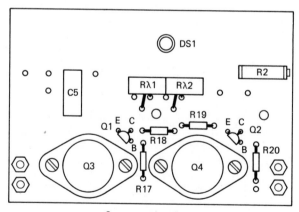

Component locations

Secure Rλ1 and Rλ2 to PC board with epoxy

(a)

Figure 9.29   Scanner driver fabrication. (A) Component
locations on E3, component side.

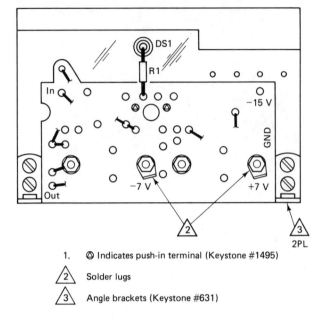

1.   ⊘ Indicates push-in terminal (Keystone #1495)

△2   Solder lugs

△3   Angle brackets (Keystone #631)

DS1 protrudes thru the PC board. Solder DS1 shell to copper.

(b)

Figure 9.29 (B) Component locations on E3, copper side.

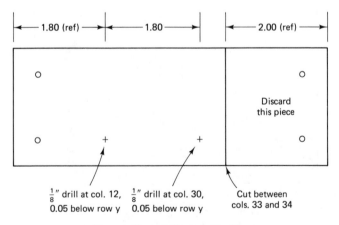

$\frac{1}{8}''$ drill at col. 12, $\frac{1}{8}''$ drill at col. 30,   Cut between
0.05 below row y   0.05 below row y   cols. 33 and 34

View of side opposite conductor side.

Dimensions are inches

Figure 9.30 Pattern for cutting and drilling E4.

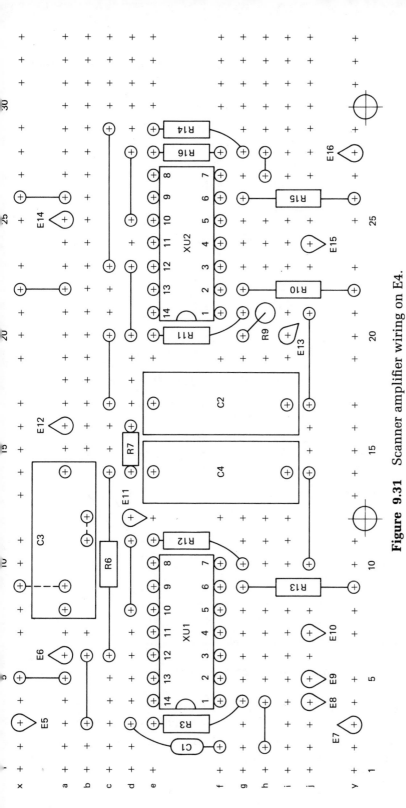

Figure 9.31 Scanner amplifier wiring on E4.

113

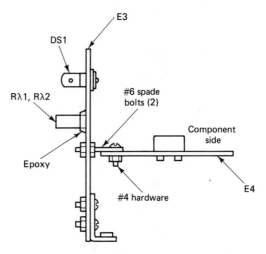

Many components are omitted for clarity

(a)

Figure 9.32   (A) View showing how E4 mounts to E3.

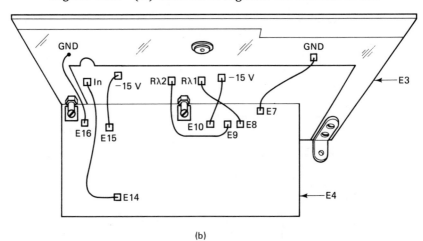

(b)

Figure 9.32   (B) Connections between E3 and E4.

both of these terminal strips to the saddle and chassis as shown in Fig. 9.37.

Connect wires to speaker and connect power supply wires to scanner driver. If you have a connector, use it here (at E1). Connect wires to PHASE and DRIVE controls as shown in Fig. 9.38 and 9.39. Connect J1 to E11. Add a 3-wire #22 power cord

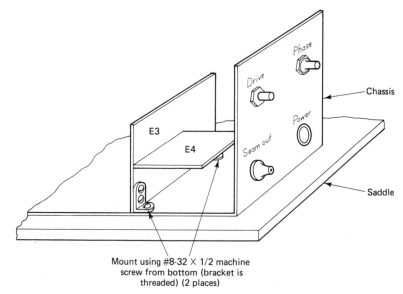

Mount using #8-32 × 1/2 machine
screw from bottom (bracket is
threaded) (2 places)

Figure 9.33   View showing how E3 (with E4) mounts to chassis and saddle.

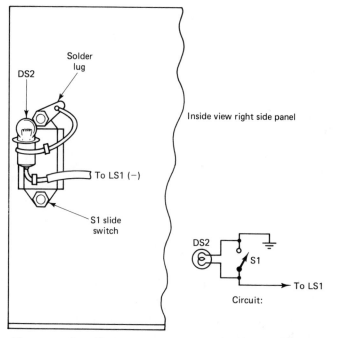

Figure 9.34   How to mount S1 and DS2 to chassis.

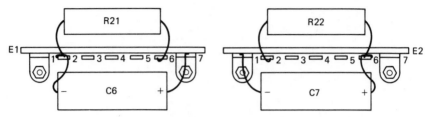

Figure 9.35 Wiring C6 and R21 to E1.   **Figure 9.36** Wiring C7 and R22 to E2.

To front panel

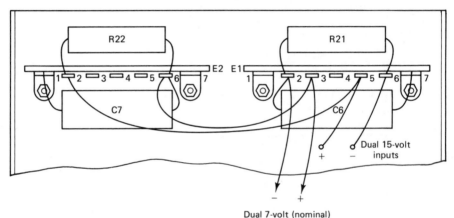

Dual 7-volt (nominal) outputs to scanner driver

Figure 9.37 View showing how E1 and E2 mount to chassis.

to bring in ±15 volt power and ground. Make this about three feet long, and terminate it with P1 as shown in Fig. 9.40. Connect other wires as required.

9.5 COMPLETING THE ASSEMBLY

Before the 7-Hz oscillator can oscillate, a shutter must be added to the end of yoke piece C as diagrammed in Fig. 9.41. Be sure the pendulum is centered when at rest. Cut the shutter from stiff black paper and fasten with white glue. Once added, oscillate the scanner

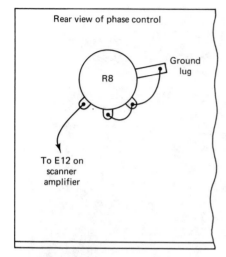

Figure 9.38 Connections to phase control.

Figure 9.39 Connections to drive control.

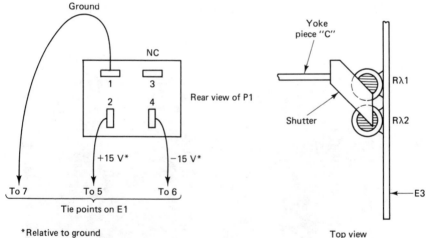

Figure 9.40 Connections to power plug P1.

Figure 9.41 View showing how shutter is cut and positioned.

pendulum manually to be sure the two photocells are eclipsed differentially and that the mechanism is clear, does not bind, and is generally free to oscillate. In rest position, the shutter

Figure 9.42   Rear view of the parallax adapter without cover.

Figure 9.43   View showing left side panel.

Figure 9.44 View showing right side panel.

should eclipse about half of each photocell as seen by the exciter lamp. Adjust the shutter outline and position to obtain these conditions.

The fiber-optic bundle can be added now or later. It is not required for normal operation. A support for the bundle at the lamp end can be fashioned from a piece of brass tubing. The front-panel end can be terminated in a banana jack. A dab of red nail polish can be added to the lamp end. Tack the bundle along its length to the saddle with silicone rubber.

You can add the half-wave retarder (the slit material) to the end of the pendulum at this time, but it is recommended that you wait until after the bugs have been eliminated from the parallax

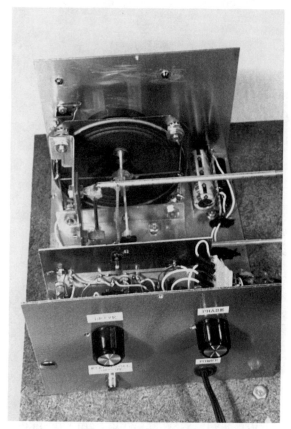

Figure 9.45 View above from left side.

adapter. That way you will be able to see how much the added mass of the slit reduces the operating frequency. The procedure for adding the slit is given in Appendix C.

Figures 9.42 through 9.47 show how your parallax adapter should look.

9.6 TIME FOR A TEST TRIAL!

When you are sure all connections have been made and there are no short circuits, you can apply plus and minus 15-V power.Ch-

Figure 9.46 View showing inside front panel.

ances are that the pendulum won't operate initially, as some adjustment is required.

Check to see that the exciter lamp lights. See that there are no hot spots or smoke. Watch the speaker cone as you connect and disconnect power. If it pulls in or out, the PWR AMP has power.

The following tests and adjustments must be performed with subdued ambient light, and with no direct light falling on the photocells other than that from the exciter lamp.

Set S1 to the SET position. With power applied, DS2 ADJ IND should light. If it does, adjust R2 BAL ADJ so that it goes out. If you cannot obtain this adjustment, this means that the shutter is not properly centered (assuming the circuit is correctly wired). Proceed by setting R2 to mid-range, and add a small wing to one side or the other of the shutter as required to cause DS2 to go out.Stick that wing temporarily in place on the shutter. (You will want to cut and attach a new shutter later to the new configuration.)

Adjust R2 so that DS2 goes out. At this point the pendulum should want to oscillate. Give it a nudge. Use the DRIVE control

Figure 9.47   Bottom view from rear.

in conjunction with S1 to set the amplitude of pendulum oscillation to 6-1/2 to 7 inches peak-to-peak.[1]

Allow the pendulum to run several hours at a time until it has accumulated 10 to 20 hours of running time to break it in. Readjust R2 as needed. It should now be self-starting.

During normal operation with S1 on SET, DS2 will flicker. With R2 properly set, it should flicker at a nominal 14-Hz rate (rather than half that rate). This indicates that both halves of the PWR AMP are working. This setting of R2 should correspond closely to its setting required to extinguish DS2 when the pendulum is centered and not oscillating. You can

---

[1]You may not be able to obtain this much swing initially. After a break-in period, the amplitude will increase.

restrain it manually to obtain this condition without fear of damaging it.

During normal operation, some components will become warm to the touch. These are R21, R22, Q3, Q4, and LS1.

With the oscillator operating, use your oscilloscope to be sure you have a 7-Hz sinewave present at J1.

Set this assembly aside and begin work on the parallax processor.

TABLE 9.1 RESISTORS, CAPACITORS, SEMICONDUCTORS

Designator	Description*
R1	100 Ω, 1/2 W
R2	10k, 10-turn trimmer pot, Mouser 32NA401
R3, R9, R15	4700 Ω
R4	5k pot, Mouser 31VA305
R5	3300 Ω
R6	100k
R7	1800 Ω
R8	1k pot, Mouser 31VA301
R10, R17, R20	33k
R11	150k
R12	220k
R13	2200 Ω
R14, R16	10k
R18, R19	820 Ω
R21, R22	33 Ω, 10 W
C1	22-100 pF ceramic
C2–C5	1-μF metalized polyester, Mouser 581-MC105K1H
C6, C7	3300-μF, 16V electrolytic
Q1	2N3417
Q2	2N3702
Q3	2N3055
Q4	2N4908, HEP 248, or MJ 2955; Hanifin
U1, U2	uA747

*Resistors are 1/4 W, 5% unless otherwise stated.
 k means kΩ here and in text.

TABLE 9.2 OTHER PARTS

Designator	Description	Suggested Source and Part No.	
DS1	Incandescent lamp, #330	Hanifin	
DS2	Incandescent lamp, #PR-3	Radio Shack	
E1, E2	Terminal strip, 2-3/8" mtg ctrs	Mouser	15TS007
(E3)	Printed circuit board kit	Radio Shack	276-1576
E4	Experimenter printed circuit board	Radio Shack	276-170
E5–E16	Terminals	Keystone	1495
J1	Panel mounting RCA-type phono jack	Radio Shack	
LS1	4" square-frame speaker, 8 Ω, 3 W	Mouser	25SP004
MP2	Steel angle bracket, legs 3-1/2 × 3-1/2"	Local hardware store	
MP4, MP5	Angle bracket	Keystone	631
P1	Jones-type plug	Hanifin	P-3304(CCT)
RΛ1, RΛ2	Cadmium-sulfide photocell	Radio Shack	276-116A
S1	SPST slide switch	Radio Shack	
XU1, XU2	14-pin DIP socket	Radio Shack	
	Interlocking chassis box, 8 × 6 × 4-1/2	Mouser	537-146-P
	Spade bolts, 6-32, 2 rqd	Keystone	1259
	Spade bolts, 8-32, 2 rqd	Keystone	1260

TABLE 9.3 MATERIALS

Part	Make From
Saddle	1/4-inch thick hardboard, 11 × 19″
Panoply, MP1, MP3	14-gauge aluminum sheets, 1 piece 11-1/8″ × 9″ 1 piece 9″ × 17″
Right-angle brackets	20-gauge aluminum sheets 2 pieces 3-1/2″ × 4-1/4″
Saddletrees	1/8 × 3/4″ aluminum strap, 2 pieces 20″ long
Pilot light	Fiber-optic bundle, 18″ long
Slit	Half-wave retarder, source: see text 1 piece 0.10 × 3.50″ see text*
Yoke	Perfboard. epoxy glass, 1/16″ thick, 1 piece, 0.2 × 0.2 grid, 4″ × 4″ 1 piece, 0.1 × 0.1 grid, 3.4″ × 0.4″
Pendulum	Thin-wall (1/64″) aluminum tubing, 1 piece 5/32″ O.D., 1 ft long 1 piece 1/8″ O.D., 3 ft long 1 piece 1/16″ O.D., 1 ft long
Other materials	1/2″ × 1/2″ aluminum angle, 10″ long
	Linear polarizer, Polaroid HN-32 0.03-inch thick: see text 1 piece 5″ × 9″, axis in the short direction 1 piece to fit your oscilloscope bezel, axis horizontal
	Music wire: 1 piece 3/32″ diameter × 9″ long 1 piece 1/16″ diameter, 3 ft long

*For orientation see Sec. 7.4 and Appendix C.

CHAPTER 10

NOW, THE PARALLAX PROCESSOR CONSTRUCTION DETAILS

10.1 THE PARALLAX PROCESSOR: SIMPLE AS AN AUDIO AMPLIFIER

The parallax processor is an all electronic instrument used to implement the H-parallax and V-perspective equations. It receives the s-signal from the parallax adapter and three spaceform-determining signals from an external source, such as those discussed in Part III of this book. It supplies plus and minus 15-V power to the parallax adapter and connects to the host oscilloscope by two cables: one for the horizontal deflection signal, the other for the vertical one.

TABLE 10.1   RESISTORS AND INTEGRATED CIRCUITS

Designator	Description*
R1, R20	10k, 10–20 turn trimmer pot
R2, R6	10k
R3	9.2k
R4	1k
R5, R21	4.7k
R6	10k
R7	10k trimmer pot
R8, R12, R15, R26, R33	10k pot, Mouser 31VA401
R9, R11, R13	47k
R10, R14	33k
R16, R24, R25	27k
R17, R18, R19, R23	22k
R22, R27, R30	3.3k
R28, R31	220k
R29	100k trimmer pot
R32	100k pot, Mouser 31VA501
R33, R34	2k
U1, U3, U5	uA741
U2, U4	multiplier/divider, Burr-Brown 4213†

*Fixed resistors are 1/4 W, 5% unless otherwise stated.

k means kΩ, here and in text.

†You may want to have an extra multiplier/divider on hand to make the sphere and hyperbolic paraboloid generators described later.

Description of the parallax adapter (given in the previous chapter) is complicated by two requirements: It has a mechanical moving pendulum, and it must have a particular physical position relative to the host oscilloscope. By contrast, the parallax processor has neither of those requirements. It is completely electronic and stands by itself. It is connected to the other subsystems only by electrical cables. Thus, there is much more leeway with its physical layout. Even so, sufficient detail is

TABLE 10.2 OTHER PARTS

Designator	Description	Suggested Source and Part No.
DS1	115-volt pilot light	Radio Shack 276-170
E1	Experimenter printed circuit board	
E2	Terminal strip	Mouser 15TS007
E3–E26	Terminals	Keystone 1495
F1	Fuse, 1/2A Slo-Blo	
J1	Jones-type connector	Hanifin S-3304(AB)
J2, J3	BNC connector, UG-1094/U	
J4	Panel mounting RCA-type phono jack	
J5, J6, J7	Binding post, red	
J8, J9, J10	Binding post, black	
MP1	"Console" cabinet, 11 × 8 × 3-1/2	Mouser 537-MDC1183-B1
PS1	±15-volt power supply	Albia PSB-203
S1	SPST 3A miniature toggle switch	
T1	Power transformer for PS1	Albia 41-01-0006
W1	115-V power cord with plug	
XF1	Panel mounting fuse holder for F1	
XU1, XU3, XU5	8-pin DIP socket	
—	T05 heat sinks, 2 rqd	Mouser 33HS502

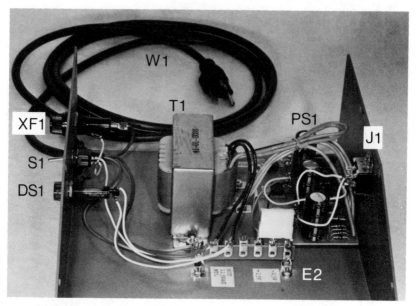

Figure 10.1 Parts location in bottom part of "console" cabinet, parallax processor.

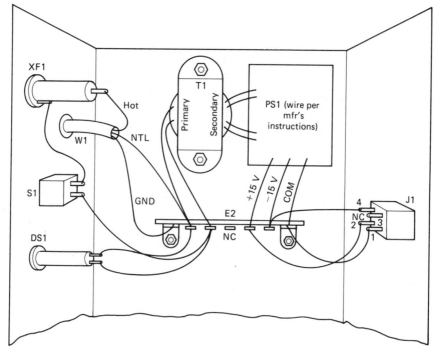

Figure 10.2 Wiring diagram, bottom of "console" cabinet.

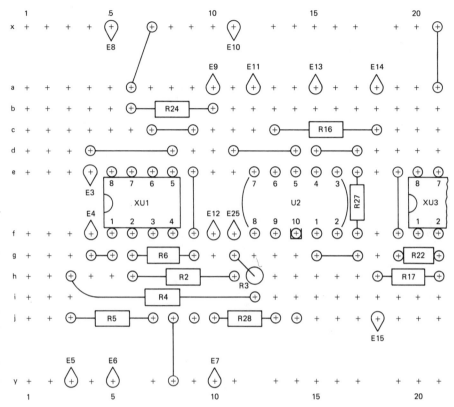

Figure 10.3 Parallax computer wiring on left half of E1.

contained in the present chapter so that the appearance and internal parts arrangement of the parallax processor shown in Fig. 8.6 can be substantially duplicated.

10.2 PARTS YOU WILL NEED AND WHERE TO GET THEM

Resistors and integrated circuits you will need are listed in Table 10.1. Other parts are listed in Table 10.2. Connector J1 (Table 10.2) mates with connector P1 of the parallax adapter. Power supply PS1 must be capable of furnishing at least 0.3 ampere on each output.

Again knobs, wire, and so on are not listed.

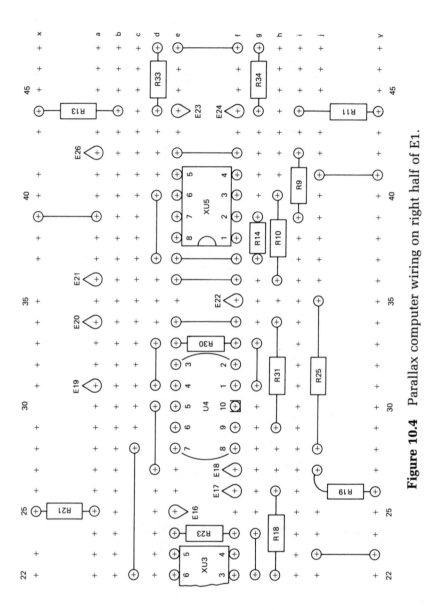

Figure 10.4 Parallax computer wiring on right half of E1.

TABLE 10.3 PARALLAX COMPUTER ADDITIONAL
WIRING

Add solid #22 insulated jumper wires between the following
indicated matrix points on E1:

Wire #	From	To
1	b 16	b 40
2	f 18	c 32
3	g 3	g 18
4	i 2	i 10
5	g 30	h 47
6	e 11	y 11
7	x 17	h 16
8	y 27	e 27
9	i 32	x 33
10	b 12	b 24
11	d 19	d 22
12	i 18	i 30
13	a 13	a 29

10.3 WIRING AND ASSEMBLING THE PARALLAX PROCESSOR

Drill holes for and mount DS1, E2, J1, PS1, S1, T1, W1, and XF1
to the bottom part of the "console" cabinet as shown in Fig. 10.1.
Wire these according to the wiring diagram shown in Fig. 10.2.
Connect the power and check that plus and minus 15 V are
both present at the proper terminals of E2. Set this subassembly
aside.

Mount and solder components to E1, except U2 and U4, as
shown in Figs. 10.3 and 10.4. Add other wires as shown in Table
10.3. Since U2 and U4 are static-sensitive devices, it is important
to understand and follow the proper procedures to prevent
damage from electrostatic discharge. An outline of some of the
requirements appears in Appendix D. Mount and solder U2 and

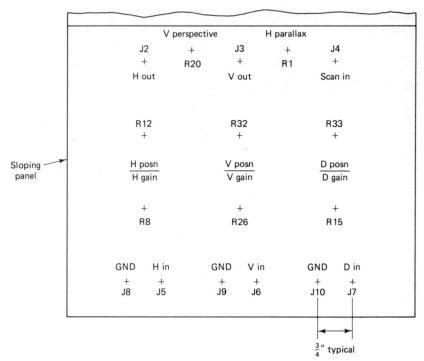

Figure 10.5 Front panel drilling layout, parallax processor.

U4 and add a heat sink to each. Be sure to mount U2 and U4 high enough above E1 to accommodate their heat sinks. This completed subassembly is the *parallax computer*.

Drill the front panel for position controls (R12 H POSN, R32 V POSN, and R33 D POSN), gain controls (R8 H GAIN, R26 V GAIN, and R15 D GAIN), connectors J2–J3, and binding posts J5–J10 as shown in Fig. 10.5.

Fashion brackets for the parallax computer, H PARALLAX pot (R1), V PERSPECTIVE pot (R20), BAL1 pot (R7), and BAL2 pot (R29) to hold them in the positions shown in Fig. 10.6. Use the panel-mounted pots to secure these brackets to the inside of the front panel. Drill adjustment holes in the front panel for R1 and R20 and mount them in such a way that they can be adjusted from outside the front panel. Mount all components and wire, following the schematic diagram, Fig. 8.8.

Figure 10.6 Inside front panel, parallax processor.

Ground the circuit to the case at a lug secured under the shell of J4 using good grounding practices to minimize the possibility of noise-producing ground loops. Connect three wires from E2 to E1 to carry plus and minus 15 V and power ground. Use a connector here if you have one.

This completes the construction of the parallax processor. Check all connections carefully to see that they are correct and that all solder connections are good with no solder bridges or other shorts. Now assemble the cabinet.

CHAPTER 11

YOUR
FIRST TEST
RUN!

11.1 IT'S FINAL ASSEMBLY TIME

Locate the cover portion of the chassis box for the parallax adapter. Cut an I-shaped slot in the front of it to accommodate the pendulum. Cut this with a nibbler to the pattern shown in Fig. 11.1. Drill two holes in the back of the cover, one as an access hole for adjusting R2 BAL ADJ, and the other to serve as a window for DS2. Add a piece of colored plastic to the inside of this window.

Paint that portion of the front surface of E3 above $R\lambda1$ and $R\lambda2$ with flat black paint to minimize reflections from that

137

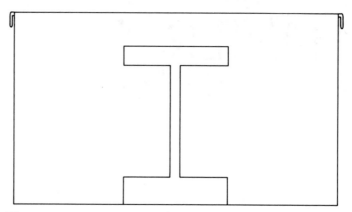

Figure 11.1   Front view of cover portion of chassis box showing I-shaped slot for scanner pendulum.

surface. You may also want to fashion a light shield for the optoelectronic portion (including Rλ1, Rλ2, and DS1) from stiff, dull, black paper. This is optional, but it will allow the pendulum to be operated in normal ambient lighting with the cover of the chassis box removed. Glue black paper to the inside top of the cover or spray it with flat black enamel if you do not use the light shield. Position the cover in place on the chassis and secure it with the two screws provided.

Spray the inside surfaces of the panoply with flat black enamel. Remove the two pieces that form the panoply before doing this, as this is probably easier than masking the remaining portions of the parallax adapter.

Add knobs to all controls. Add the slit to the slit carrier on the pendulum if you have not already done so, using the procedure given in Appendix C. Cut two pieces of linear polarizer to the dimensions given in Table 9.3. Attach the smaller one to the oscilloscope bezel. Put it between the CRT and the graticule if possible. In this way, the graticule illumination will be steady, not scanned by the slit. Add the larger polarizer to the panoply, attaching it to the rear of the front panel with silicone rubber as shown in Fig. 11.2.

Add labels to panels adjacent to all controls and connectors.

Set the parallax adapter in position atop the host oscilloscope. Plug P1 into J1 of the parallax processor.

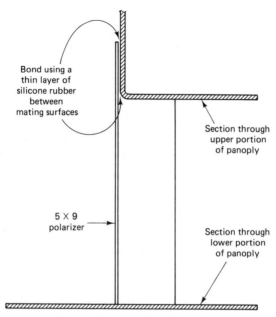

Figure 11.2   View showing attachment of polarizer to panoply.

Connect a shielded cable from J1 on the parallax adapter to J4 on the parallax processor. Connect two coax cables from the parallax processor to the host oscilloscope; one from J2 to the horizontal input of the oscilloscope, the other from J3 to the vertical input of the oscilloscope. The oscilloscope controls must be set for XY operation in this application.

11.2 CRANKING UP AND ADJUSTMENT INSTRUCTIONS

Plug in the power cord W1 of the parallax processor to 115 V 50 to 60 Hz power. Switch on the power (S1). The pendulum should now run. Trim BAL ADJ if necessary. Switch the power off and unplug W1.

Remove the front panel from the parallax processor. Plug in and switch on. With the scanner pendulum running but no inputs to J5, J6, and J7 (no deflection inputs), set BAL1 pot so that the average voltage at pin 6 of U1 is zero. Set BAL2 pot so that the average voltage at the horizontal output (J2) is zero with H POSN set to midway position. Switch the power off, unplug W1, and replace the front panel.

Set the oscilloscope controls as follows:

1. Set mode for XY operation.
2. Set input deflection circuits for dc coupling.
3. Set horizontal channel for 500 mV/cm sensitivity.
4. Set vertical channel for 1 V/cm sensitivity.
5. Set position controls to center the spot.

11.3 FINE TUNING THE OPERATION

Your parallax adapter, parallax processor, and oscilloscope should still be interconnected and ready to serve as a parallactiscope. Dim or extinguish overhead lights. Apply power and check that the scanner pendulum is operating properly and that there is a spot on the CRT. Set GAIN controls on the parallax processor fully CCW (counterclockwise). Center the spot using POSN controls on the parallax processor. If there are two spots, turn the PHASE control on the parallax adapter until they fuse into a single spot. There may remain some ghosting due to the persistence of the CRT phosphor and due to possible misorientation of the two linear polarizers. Correct this latter condition by re-orienting if it exists.

Exercise all three POSN controls, noting their action. Turning the H POSN control CW should cause the spot to move to the right.[1] Turning the V POSN control CW should cause the spot to move upward. Turning the D POSN control CW should cause the spot to appear to move toward you. Set the oscilloscope's horizontal position control so that the spot appears to move perpendicularly to the CRT screen as you turn the D POSN control. Now center the spot using the H POSN control.

---

[1]This and following descriptions assume that your host oscilloscope has positive-deflection directions rightward (horizontal channel) and upward (vertical channel). Check this with a +5 V source.

Apply a sinewave of about 5 kHz to V IN of the parallax processor. Apply a synchronized sawtooth wave to H IN and adjust controls to give a standard sine waveform display. Adjust H PARALLAX and V PERSPECTIVE so that as you turn the D POSN control, the pattern moves "in" or "out" without distorting. H PARALLAX controls pattern growing or shrinking in the horizontal direction. V PERSPECTIVE controls this in the vertical direction. Multiscopic photos of these displays are shown; Figures 11.3, 11.4, and 11.5 show the waveform at the screen position, behind the screen, and in front of the screen, respectively.

Multiscopic photos consist of a series of display photos taken from different horizontal angles. The photos are then mounted side-by-side so that they can be viewed in pairs stereoscopically. This shows the stereoscopic depth that occurs naturally in the live parallactiscope displays. By moving your

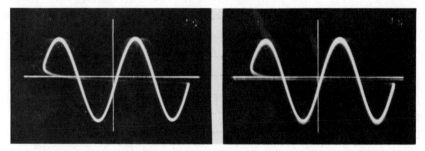

Figure 11.3 Multiscopic views of a holoform oscillogram (parallactiscope display. Subject is a sinewave at the screen position. This is a display without depth.

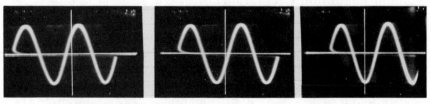

Figure 11.4 Multiscopic views of a sinewave behind the screen position. View stereoscopically in pairs (uncrossed disparity). Camera viewing angle was changed 10 degrees between adjacent photos using the fixture described in Appendix E.

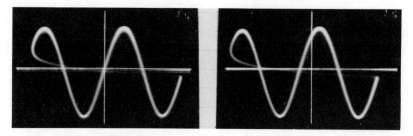

Figure 11.5 Multiscopic views of a sinewave in front of
the screen.

view to the next adjacent pair of photos, you can get something
of the feel for the movement parallax that occurs naturally in the
live displays. These photos were taken using the apparatus
described in Appendix E.

The use of multiscopic photos is only one way to record or
store the image information. Another way is to convert the three
signals to digital form and store them in memory. Then when
you want to view these patterns, they can be played back into the
parallactiscope via digital-to-analog converters.

11.4 IF IT'S "BROKE," FIX IT!

This section presents some procedures that were indispensible
during the design phases of parallactiscope development. These
procedures may prove helpful to you if you experience difficulty.

Scanner Pendulum Oscillation

To cause the pendulum to oscillate, gains and phases around the
feedback loop containing the pendulum must be correct. To
check these, break the circuit at the output of U2B (pin 10). Insert
an external sinewave to the scanner driver power amplifier
(junction of R18 and R19). Adjust its frequency (nominally 7 Hz)
and amplitude to cause the pendulum to oscillate at resonance.

Display this signal versus the output of U2B on an XY
oscilloscope display. If the resulting pattern has a loop, this
means there is a phase error. If the pattern forms a straight line

but is not at a slope of $+1$[2] when both deflection sensitivities are equal, this means there is a gain error. When you can obtain a straight line whose slope is $+1$, the circuit is right, and the connection from U2B pin 10 to the junction of R18 and R19 can be restored (removing the external signal source) and the pendulum should continue to oscillate, the circuit now being self excited.

If the slope of the line is negative, check that the speaker connections are not reversed. For proper phasing, the speaker terminal marked with a red dot must connect to the power amplifier. The red dot means that a positive voltage there causes the speaker cone to move out, away from the magnet. This action can be checked by first disconnecting the speaker from the power amplifier, then connecting a size C or D flashlight cell to the speaker while watching the cone to see which way it moves. If your speaker does not have a red dot, add one at this time.

Scanner Signal (S-Signal)

The s-signal must be adjustable by the PHASE control to be exactly in-phase with the scanner pendulum motion. This is necessary so that the image seen for leftward slit motion will coincide exactly with that seen for rightward slit motion. The operation of this circuit can be checked as follows.

Remove the two coax cables from the oscilloscope. Switch to TY mode, with time base set to 20 microseconds per centimeter or faster. Inject the s-signal output from the parallax adapter into the vertical input of the oscilloscope. Set the pendulum in operation in position on the oscilloscope. The pattern on the CRT screen should be a horizontal line that bobs up and down at the pendulum frequency. When viewed through the panoply, this should appear as a loop or a diagonal line with negative slope.[3]

Turn the PHASE control. If you can adjust for a line (as opposed to a loop), then this circuit is correct. In this display, the s-signal is providing the vertical deflection and the slit is

[2]See Footnote 1 of this chapter.
[3]See Footnote 1 of this chapter.

providing the horizontal deflection. If these are in-phase, you will see a straight line—the desired condition.

If you are unable to adjust the PHASE control (R8) for a line, this means resistor R7 needs trimming. To reduce its value, parallel it with an 1800-Ω resistor. To increase its value, connect a 1000-Ω resistor in series with it at R8.

Measuring Scanner Pendulum Frequency

You may wish to know accurately the frequency of the pendulum. If you have sophisticated instrumentation available, this presents no problem. Such instruments as sinusoidal frequency counters and accurately calibrated stroboscopes are certainly appropriate.

If you do not have such instrumentation available, you can build the simple frequency meter described in Appendix F to give the frequency to two or more significant figures, depending on the time span over which you elect to count audible ticks. The meter provides one tick for every two cycles. You can count easily at that rate, about seven ticks every two seconds.

11.5 NEXT STEP: HOW TO USE WHAT YOU JUST BUILT

Your parallactiscope is now complete and ready to be put to use. The kinds of images you will obtain depend on the kinds of signals you inject into the three deflection inputs of the instrument.

The H-parallax and V-perspective transformations that are performed by the parallactiscope cause the proper "synthetic retinal images" to be generated so as to faithfully convey the desired 3-D spaceform information pattern to your centers of cognition. This is so regardless of the makeup of the three waveforms you use as inputs to the parallactiscope. In all cases, the information pattern you will sense upon viewing these images is governed by the rules of 3-dimensional graphing explained in the following chapter.

PART III

SIGHTS YOU'VE NEVER SEEN!

CHAPTER 12

WHAT YOU CAN LOOK AT WITH YOUR PARALLACTISCOPE TODAY

12.1 TWO BASIC CATEGORIES OF IMAGES YOU CAN SHOW

Just as two synchronized waves (periodic functions) repetitively define a 2-dimensional image in 2-D oscillography, so do three synchronized waves repetitively define a 3-dimensional image in 3-D oscillography. Such a trio of signals can come from virtually any source and can have virtually any composition within the voltage and frequency ranges of the display device.

147

The signals and images of 3-D oscillography can be categorized as follows: those used to analyze the operation of a device or system, and those used to synthesize a particular desired space curve or surface. There is an overlap in the two categories. For example, the surface that is characteristic of the operation of an analog multiplier is analytic when its purpose is to show the operation of the multiplier, and synthetic when it is to show a particular mathematical surface—a hyperbolic paraboloid.

12.2 ANALYZING SIGNALS WITH YOUR PARALLACTISCOPE

The rules of *2-dimensional parametric graphing* (with time as the parameter) determine the curve generated by two functions plotted orthogonally to each other. This can be expressed mathematically as follows:

$$x = f_1(t)$$
$$y = f_2(t)$$ (12.1)

Each point of the curve is determined by substituting a particular value of t (time) into the two functions. As time flows, the point moves in the xy plane, tracing out the curve defined by the two functions.

In 2-D oscillography, a sawtooth wave or ramp—a "time base" signal—is used on the horizontal axis (here represented by x), and the signal to be analyzed is used on the vertical axis. This is the TY mode. In the XY mode, two external signals are used.

In the same way, the rules of *3-dimensional parametric graphing* apply to parallactiscope images. In general, a 3-dimensional spaceform is determined by three functions plotted orthogonally to each other. This can be expressed mathematically as follows:

$$x = f_1(t)$$
$$y = f_2(t)$$
$$z = f_3(t)$$ (12.2)

The "rules of graphing" for 3-dimensional figures are simply an extension of the rules for 2-dimensional figures. To find the x, y, z coordinates of the plotted point at a particular instant of time, substitute that value of t into the three functions and evaluate them. Then as time flows, the point moves, tracing out the curve or surface defined by the three functions.

While this is easy to describe, it is a tedious process when done manually (not to mention the general unavailability of "3-dimensional graph paper"). It is done automatically for you if you have the three functions in the form of electrical signals and if you have a parallactiscope to "play them back" into.

Take a lesson from 2-D oscillography, and extend by one the number of signals you can cause to be plotted against another. Or you can use only two signals to present waveforms or other 2-D spaceforms at unusual angles.

In 3-D oscillography, 3-dimensional "waveforms" are easy to display. Such an image might present voltage versus current versus time for a particular device. In this case, one of the signals is a sawtooth time base. Two sawtooth waves can be used to produce a horizontal raster, with a third signal used to present an elevation plot on the vertical axis. Or all three inputs can be signals derived from a device or system.

12.3 GENERATING IMAGES YOU WOULD LIKE TO SEE

If you desire to synthesize a particular twisted curve or surface, you must first determine what trio of functions defines it. You can do this by starting with either the equations or graphs of the functions.

If you have mathematical expressions for the three functions of Eqs. (12.2), then you can devise a simple special-purpose "analog computer" to generate the signals. With the current availability of operational amplifiers and other analog building blocks, this is much easier to do today than it was just a few years ago.

If you do not have mathematical expressions for the three functions, you must determine them graphically. Then you might generate them using either analog or digital techniques. In any case, ingenuity helps. One thing working in your favor is the fact that *there will always be more than one trio of functions*

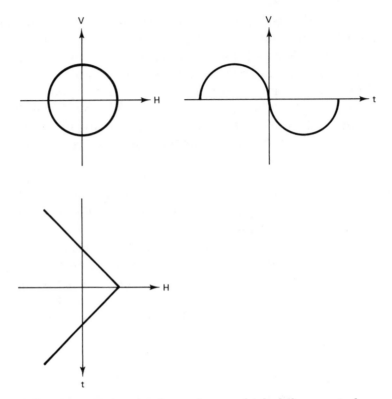

Figure 12.1   A pair of waveforms which defines a circle.

defining a given spaceform. As unlikely as this might seem, consider that a circle can be described by a pair of quadrature sinusoids and by the functions graphed in Fig. 12.1. The difference is that the circles are traced out at different rates. With quadrature sinusoids the rate is constant; with the functions of Fig. 12.1 it is not.

12.4 COMMENT

As you explore the uses of your parallactiscope, one of the things you may notice is that the amount of depth can be increased by moving the parallax adapter forward (toward you) on the host oscilloscope. Another thing you may notice is that there is a relationship between slit-scan amplitude and image brightness.

Using the full seven inches of peak-to-peak slit-scan amplitude, you should be able to move as much as 1/4 piradian (45°)
either way from center (the "head-on" position), and still see
nearly all of the display visual field. With reduced slit-scan
amplitude, this maximum observer–movement angle will be
reduced but image brightness will be increased. You can reduce
the slit-scan amplitude by use of the DRIVE control.

In the following two chapters, examples are given of 3-D
spaceforms and ways to produce them on your parallactiscope.
In none of these is the use of a digital computer required. The
oscilloscope is basically an analog device—the parallactiscope
carries on this tradition.

It has been said that in the future everything will be
digitized! Even so, analog ideas and analog approaches to
problem-solving are still alive and well today.

CHAPTER 13

HOW TO SEE
SIGNAL PATTERNS
IN 3-D FOR
ANALYSIS

13.1 SINUSOIDAL WAVEFORMS SEEN END-ON!

A common figure of 2-D oscillography is the sinusoidal waveform. This is produced by using a sawtooth wave on the horizontal input and a sinusoidal wave on the vertical input. This kind of display can also be produced on the parallactiscope as was noted in Section 11.3 where this classic figure was used to aid in setting the H PARALLAX and V PERSPECTIVE adjustments.

With that display, the sawtooth wave is applied to the H INPUT. If it is applied to the D INPUT instead, you will see the waveform on a vertical plane perpendicular to the screen! Figure 13.1 shows multiscopic views of such a pattern.

In the same way that the sinusoidal waveform on end is produced, any waveform can be presented end-on. Multiscopic views of a triangular waveform presented that way appear in Fig. 13.2.

13.2 DISPLAYING THE ABSTRACT SURFACES CHARACTERISTIC OF ELECTRONIC DEVICES

Some electronic devices are fully characterized by two variables. A diode is an example of this. Others require three or more variables to characterize them. The parallactiscope is well suited to the display of such characteristic curves and surfaces. Even diode characteristics can be displayed to advantage on the parallactiscope.

A diode characteristic can be displayed as a 3-dimensional "waveform" by plotting voltage versus current versus time. A circuit diagram for accomplishing this is shown in Fig. 13.3. The resulting "waveform" is shown in the multiscopic views of Fig. 13.4. Alternatively, this can be presented as the bent surface shown in Fig. 13.5.

Transistor curve tracers are special-purpose 2-D oscilloscopes that display a family of 2-D curves representing the 3-D surface of, say, collector current versus collector-emitter voltage versus base current. Such characteristics can be shown on the parallactiscope in the same form, or as an actual surface.

In principle, at least, devices requiring four variables to characterize them can be analyzed as families of 3-D surfaces. This is analogous to the use of families of 2-D curves to analyze 3-D characteristics of transistors in the transistor curve tracer.

13.3 SEEING THE CHARACTERISTIC SADDLE-SHAPED SURFACE OF AN ANALOG MULTIPLIER

Analog multiplier operation can be checked by visual examination of the surface that is characteristic of its operation. The shape of this surface is independent of the signals fed to

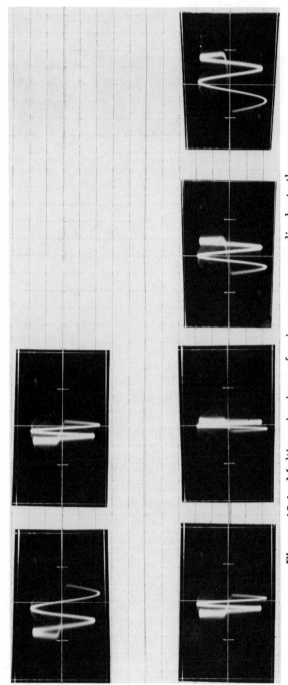

Figure 13.1 Multiscopic views of a sinewave perpendicular to the screen.

155

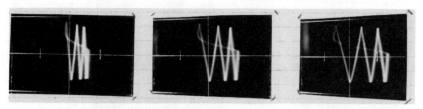

Figure 13.2 Multiscopic views of a triangular waveform perpendicular to the screen.

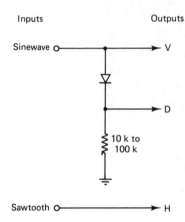

Figure 13.3 Circuit for producing the diode characteristic. Use any diode.

it, as long as they are within its ratings in frequency and amplitude.

A 4-quadrant multiplier (we used a Burr-Brown 4213) produces a symmetrical saddle-shaped surface: a hyperbolic paraboloid. Any deviation from that surface shows that the multiplier is bad, or that it is being operated outside it ratings. This is so regardless of the signals used on its two inputs.

The surface is obtained by looking at the output on the vertical axis plotted against the two inputs on the horizontal and depth axes. Figure 13.6 shows multiscopic views of this surface produced in this manner. The circuit used is shown in Fig. 13.7.

To prove that the surface shown in Fig. 13.6 is really a hyperbolic paraboloid, start with a standard representation of that surface

$$x^2 - y^2 = 2pz \qquad (13.1)$$

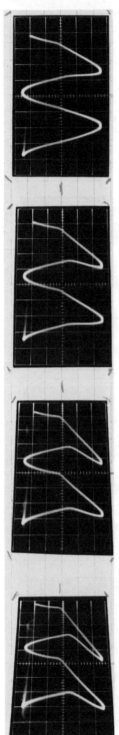

Figure 13.4 Multiscopic views of voltage versus current versus time for a diode; sinewave input.

Figure 13.5 Same as Figure 13.4 but displayed as a surface.

157

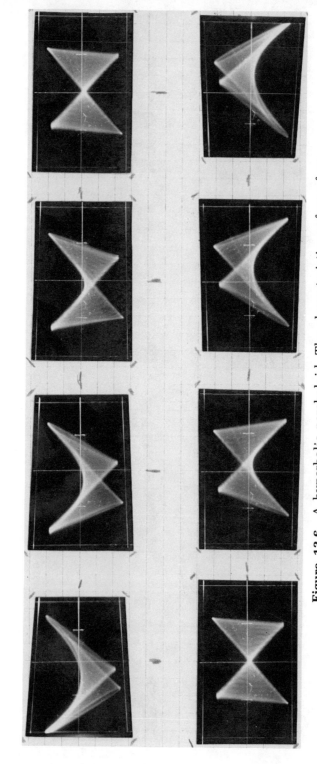

Figure 13.6 A hyperbolic paraboloid. The characteristic surface of an analog multiplier.

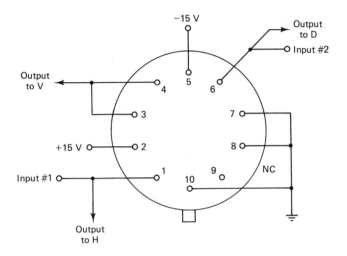

Bottom view of Burr–Brown 4213

Figure 13.7 Circuit for producing the multiplier characteristic surface.

and rotate it 1/4 piradian (45°) about the z-axis. The rotated coordinates are x′, y′, z with

$$x = x'/2^{\frac{1}{2}} + y'/2^{\frac{1}{2}}$$
$$y = y'/2^{\frac{1}{2}} - x'/2^{\frac{1}{2}}$$ **(13.2)**

A little algebraic manipulation now shows that Eq. (13.1) becomes

$$z = x'y'/p$$ **(13.3)**

where p is a constant.

Equation (13.3) describes a hyperbolic paraboloid. Since Eq. (13.3) also describes the action of an analog multiplier, with x′ and y′ the multiplier inputs and z the output, this proves that the characteristic surface of an analog multiplier is a hyperbolic paraboloid.

Hyperbolic paraboloids are ruled surfaces, but they are not "developable" surfaces. A ruled surface is one that can be completely described by straight lines. Those straight lines are called *rulings*. A developable surface is one that can be fit onto a plane without stretching; it can be cut and bent. A cylinder is an example of a developable surface and a ruled surface.

If your host oscilloscope has an external intensity-modulation (Z-axis) input, this can be used to advantage when displaying figures such as the hyperbolic paraboloid. Simply inject the D-axis signal into it. Phasing must be such that the more remote portions of the figure can be dimmed or blanked. In this way, displays can be given a solid appearance as opposed to a transparent one. Such a procedure does not accomplish a rigorous solution to the "hidden-line problem," but it does give a satisfactory approximation of it.

13.4 A MEDICAL APPLICATION: THE VECTORCARDIOGRAM

Oscilloscopes are being used more and more in medicine. One such use is in the display of EKGs (electrocardiograms). These are 2-D entities, and so a 2-D oscilloscope suffices for their display. But there is such a thing as a 3-D EKG—a *vectorcardiogram* (VCG). Without a 3-D oscilloscope, VCGs are difficult to use because it is necessary to construct physical wire models of these twisted curves to view them. Because they are difficult to use, it is doubtful that their full potential has been realized.

VCGs are obtained by mixing (matrixing) the electrical signals obtained from several strategically placed body electrodes. From this, three pulses are obtained from heart action that describe a periodically recurring loop in 3-D space. While such signals cannot be viewed directly with the parallactiscope (for reasons to be explained presently), they can be recorded on magnetic tape and subsequently played back at a faster speed into the parallactiscope. The reason for this mode of operation is as follows.

In the parallactiscope, the slit scans at a rate of about fourteen scans per second. As a result, signals within a band of frequencies centered logarithmically on 14 Hz will produce visual beats or Moiré patterns because of interference with the slit scan. To avoid this *Moiré band*, signal frequencies must be greater than about 500 Hz to avoid being broken up—except that the dc components of signals are faithfully depicted.

Figure 13.8 Multiscopic views of a simulated vectorcardiogram (VCG).

Perfectly repetitive signals having frequencies within the Moiré band can be displayed directly on the parallactiscope for photographing monoscopically, stereoscopically, or multiscopically. This is practical with time-exposure photography because gaps occurring in one scan tend to fill in on subsequent scans. It should be noted, however, that raw EKG and VCG signals are not perfectly repetitive.

Multiscopic views of a simulated VCG are shown in Fig. 13.8. These photos may give some indication of the usefulness of the parallactiscope to display VCGs.

CHAPTER 14

SYNTHESIZING PARTICULAR 3-D SPACEFORMS

14.1 PROGRESSING FROM SINEWAVE TO HELIX

Jules Lissajous showed that two sinusoids can be used to produce looping 2-D patterns that we now call *Lissajous figures*. It is easy to expand this concept to include 3-D patterns by simply including another sinusoid. Such a 3-D Lissajous figure is shown in the multiscopic photos in Fig. 14.1.

If instead of plotting one sinusoid against another sinusoid, it is plotted against a sawtooth wave; the classic sinusoidal waveform display is obtained as discussed in the previous chapter. Similarly, two sinusoids can be plotted against a

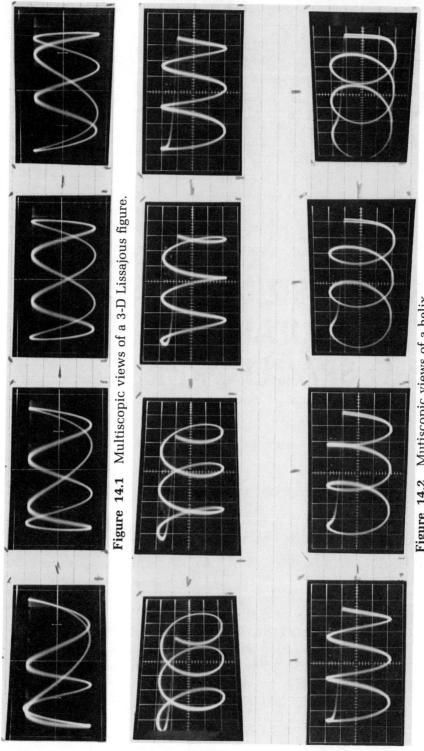

Figure 14.1 Multiscopic views of a 3-D Lissajous figure.

Figure 14.2 Mutiscopic views of a helix.

sawtooth wave. The result is a *helix*. A helix produced in this manner is shown in the multiscopic sequence in Fig. 14.2.

The helix of Fig. 14.2 is produced with the circuit shown in Fig. 14.3. In this case, the two sinusoids are in phase quadrature and the sawtooth wave has a frequency that is a third that of the sinusoidal waves. This helix is partly in front of the screen and partly behind it. Figure 14.4 shows a helix entirely in front of the screen; Fig. 14.5 shows one entirely behind it.

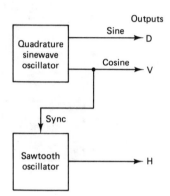

Figure 14.3   Circuit for producing helix.

The helix in Fig. 14.6 is behind the screen, and is partially eclipsed by the edge of the screen. Note that you can see the eclipsed portion by moving to the opposite side of the screen and peering around (moving to other pairs of photos, in the multiscopic presentation here); just as you would with a real object that is on the far side of a window.

It is stressed that these multiscopic photos were produced from a single parallactiscope display and that the controls were not changed between photos in a given sequence. The only thing that was changed was the camera position to simulate changes in the observer's eye position. This is true for all multiscopic photo sequences appearing in this volume.

14.2 SEEING THE DOUBLE HELIX OF DNA

The equations of a helix can be written

$$x = \sin \omega t$$
$$y = \text{saw } k\omega t \qquad\qquad (14.1)$$
$$z = \cos \omega t$$

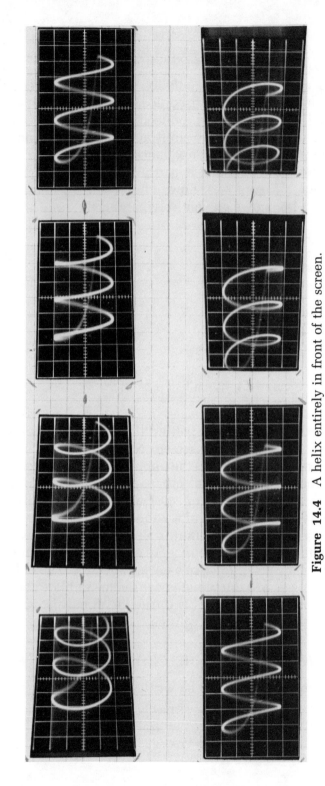

Figure 14.4 A helix entirely in front of the screen.

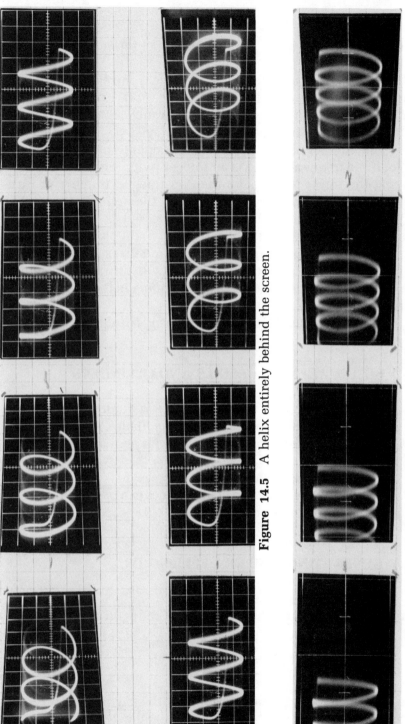

Figure 14.5 A helix entirely behind the screen.

Figure 14.6 A helix behind the screen, moved to be partly eclipsid by the screen edge.

where saw $k\omega t$ is a sawtooth wave synchronized with the sine and cosine waves. The functions in Eq. (14.1) produce a helix with axis in the y-direction. When three such signals are injected into the parallactiscope a helix is displayed.

The number of turns on the helix depends on the frequency ratio. If the frequency ratio is an integer (with the sawtooth wave having the lower frequency), then an integral number of turns equal to that frequency ratio appears on the helix. If it cannot be expressed as an integer but can be expressed as half an integer, then a double helix results.

A double helix is equivalent to two helixes interlaced. In a similar way, a frequency ratio that is a third of an integer produces a triple helix, and so on. A double helix so obtained is shown in Fig. 14.7.

The replicating DNA molecule forms a double helix.

14.3 SEEING "DENSE" HELIXES AND CYLINDERS

By further reducing the frequency of the sawtooth wave relative to that of the two quadrature sinusoids, "dense" helixes can be displayed. Such a helix having five turns is shown on end in Fig. 14.8.

By causing the sawtooth wave to be asynchronous with the two sinusoidal waves, cylindrical surfaces can be produced. One such surface is shown on end in Fig. 14.9.

14.4 GENERATING SPHERES, ELLIPSOIDS, AND CONES

Three functions that define a sphere of radius R are

$$
\begin{aligned}
x &= R \cos \theta_1 \cos \theta_2 \\
y &= R \sin \theta_1 \cos \theta_2 \\
z &= R \sin \theta_2
\end{aligned}
\tag{14.2}
$$

From Eq. (14.2) you can determine the signals required to generate a sphere or ellipsoid on the parallactiscope. They are represented by

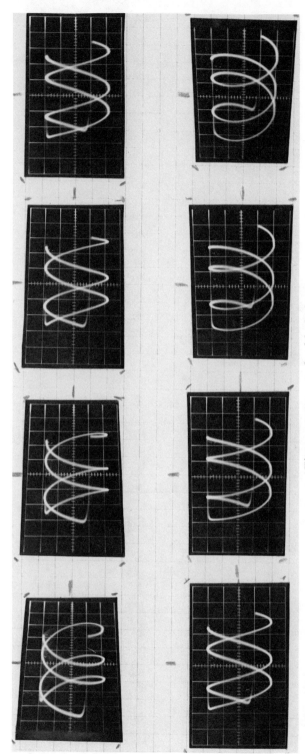

Figure 14.7 A double helix.

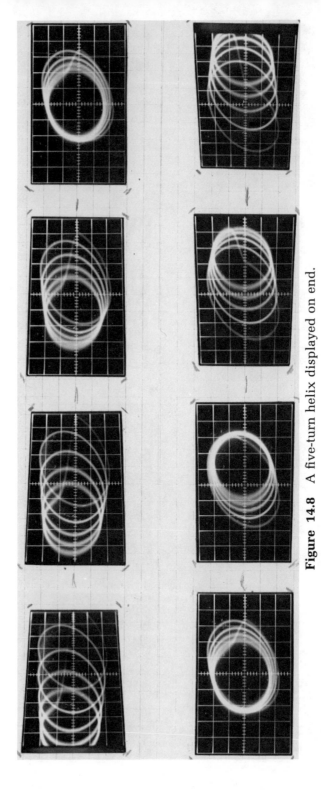

Figure 14.8 A five-turn helix displayed on end.

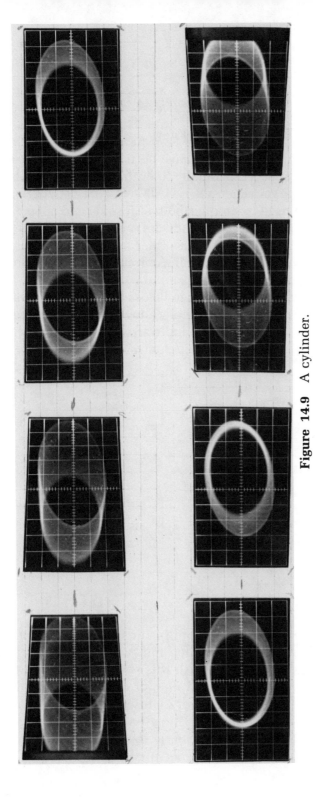

Figure 14.9 A cylinder.

$$x = \cos \omega_1 t \, \cos \omega_2 t$$
$$y = \sin \omega_1 t \, \cos \omega_2 t \qquad \textbf{(14.3)}$$
$$z = \sin \omega_2 t$$

where the amplitude factors have been omitted. The three amplitudes can be adjusted manually to desired values by means of the parallactiscope gain controls (H GAIN, V GAIN, D GAIN). Thus, Eq. (14.3) can be used to produce both ellipsoids and spheres.

If $\omega_1 >> \omega_2$, lines of latitude are described. If $\omega_2 >> \omega_1$, lines of longitude are described. In either case, the two poles of the sphere lie on the z axis. Thus if z corresponds to the H input, the axis of the figure will be horizontal.

Two analog multipliers can be used to synthesize the x and y functions of Eq. (14.3), and thereby allow you to display a spherical or ellipsoidal solid on your parallactiscope. The block diagram of this special purpose "analog computer" is shown in Fig. 14.10.

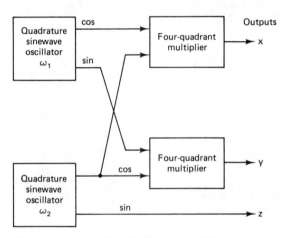

Figure 14.10 Block diagram of Eqs. (14.3).

If $\omega_1 >> \omega_2$ spheres and ellipsoids can be generated using only one 4-quadrant multiplier and two RC phase-shifting networks. A circuit to accomplish this is shown in Fig. 14.11.

In Fig. 14.11, the multiplier produces the signal $\cos \omega_1 t$ $\cos \omega_2 t$. The upper RC network subsequently produces $\sin \omega_1 t$ $\cos \omega_2 t$ from that by phase shifting the higher-frequency component (the "carrier") while leaving the lower-frequency component (the "modulation") largely unaffected. Finally, the lower

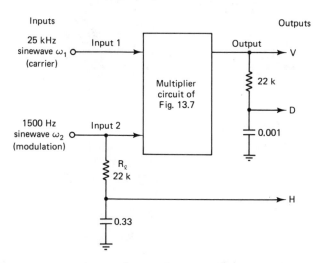

Figure 14.11   Circuit for producing spheres and ellipsoids.

RC network phase shifts one oscillator output to produce the third signal, sin $\omega_2 t$.

Multiscopic photos of displays produced using the circuit of Fig. 14.11 are shown in Figs. 14.12 to 14.15. Fig. 14.12 shows an oblate spheroid (a cushion shape).[1] Figure 14.13 shows a sphere, and Fig. 14.14 shows a prolate spheroid (football shape). If you bypass the lower RC network (let $R_\ell = 0$), a cone of two nappes results. This is shown in Fig. 14.15.

The fact that the same circuit can produce cones and spheres suggests an experiment: If shifting the phase of z by 1/2 piradian produces a sphere, and not phase shifting at all produces a cone, what will shifting it by some intermediate value do? A funny thing happens when you try this simple experiment: The result is a *conoid*—a cross between a sphere and a cone of two nappes!

14.5 CONOIDS: A CROSS BETWEEN SPHERES AND CONES!

A gradual change from a sphere to a cone can be accomplished by using the connection for a sphere (in Fig. 14.11), making R_ℓ variable, and slowly reducing the resistance R_ℓ from maximum

[1]A spheroid is a particular kind of ellipsoid.

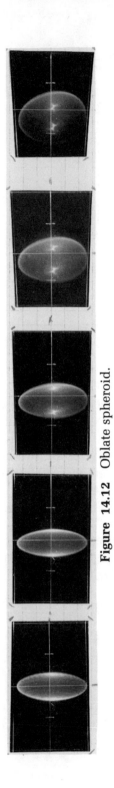

Figure 14.12 Oblate spheroid.

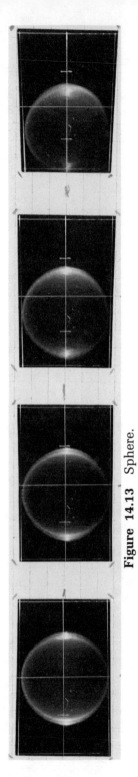

Figure 14.13 Sphere.

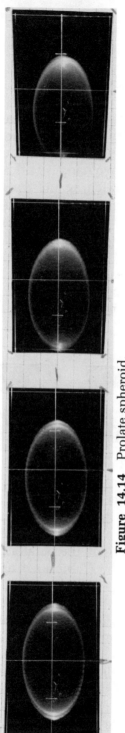

Figure 14.14 Prolate spheroid.

Figure 14.15 A cone of two nappes.

175

to zero. The metamorphosis, as observed on the parallactiscope, proceeds as follows.

Starting with a sphere, as R_ℓ is decreased, the sphere slowly splits into two spheroids as if it had two skins. The two resulting figures move apart, horizontally, *through each other* and begin to distort. Each will become one of the two nappes of the cone. The distortion consists of a horizontal stretching and flattening, and a dimpling inward at the outer poles of the two halves of the figure. This trend continues as R_ℓ is further decreased—the dimpled portions poking inward to form inside surfaces of the two nappes. When R_ℓ reaches zero, the metamorphosis is complete and the cone, in its two nappes, finally appears.

Multiscopic photos of an intermediate stage are shown in Fig. 14.16. The separate *intersecting* inner and outer skins of the cone-to-be are clearly shown there.

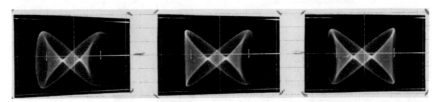

Figure 14.16   Conoid.

The figure is called a conoid because the limit of this set of figures is a cone, just as the limit of the spheroids is a sphere.

The process may be difficult to visualize from the written description just given. When seen live on the parallactiscope no words are needed. To paraphrase the Chinese, if a picture is worth a thousand words, then perhaps a parallactiscope image is worth a thousand verbal descriptions!

The above description tells of only one of the multitude of visual mathematical adventures available to you with your parallactiscope. The joy is not only in seeing the figures and how they change, but also in the ease of relating those images to their associated mathematical expressions. "Ease" because of the one-to-one correspondence of (analog) electronic block diagrams to mathematical equations. Such correspondence exists not only for the classic figures of geometry, it also exists for all conceivable equations and all conceivable block diagrams. Thus, these images are intellectually meaningful as well as being visually satisfying.[2]

Have we invented a new mathematical figure in the conoid? Have you seen reference to it in any mathematical volume? Its describing equations are

$$x = A \cos \omega_1 t \cos \omega_2 t$$
$$y = B \sin \omega_1 t \cos \omega_2 t \qquad \text{(14.4)}$$
$$z = C \sin (\omega_2 t + \phi), \ 0 < \phi < \pi/2$$

14.6 INVENTING AND GENERATING YOUR OWN 3-D CURVES AND SURFACES FOR LIVE VIEWING ON THE PARALLACTISCOPE

Both this and the previous chapter give specific examples of curves and surfaces you can display on your parallactiscope. There is virtually an infinite number of others that you can also display.

For example, consider the other classic figures of 3-dimensional geometry. Besides cylinders, cones, spheres, ellipsoids, and hyperbolic paraboloids, there are elliptic paraboloids and hyperboloids of one and two sheets.

One way to generate the signals required for mathematical figures is to express their equations in parametric form with t (or ωt) as parameter, construct the block diagram of those functions, and subsequently implement the block diagram with analog computing building blocks.

The same kind of procedure can be used to generate curves and surfaces that do not appear in the bulk of the mathematical literature, as in the case of the conoids. The opportunity and challenge to use and develop your creative abilities in this regard are both apparent and considerable.

There is an excitement about pursuing these kinds of mathematical investigations and seeing them "come alive" in three dimensions.

---

[2]For further insights into this fascinating relationship between block diagrams and equations, and hints on how to exploit it, see Ref. 7 in Appendix H.

While multiscopic photos provide some of the stereoscopic and parallactic information generated by the parallactiscope, they are no substitute for viewing the live images. One cannot capture in photographs the dynamic nature of the patterns— their immediacy, the sparkling green "fire" produced as the images are traced rapidly and repeatedly. These are visual objects floating in space. It seems that you could reach in there and grasp them.

The end purpose of this book does not reside in the multiscopic photos presented; its purpose will be fulfilled only when you have built your own parallactiscope, used it, and improved upon it.

CHAPTER 15

SUPERPARALLACTISCOPE: DESIGNING A MANUALLY-CONTROLLED IMAGE ROTATOR FOR YOUR PARALLACTISCOPE

15.1 INTRODUCTION

One of the features of the scenoscope is the capability of rotating the 3-D images by use of a manually-operated rotation control. The same kind of system can be used to advantage with the parallactiscope.

The rotator can take the form of an outboard module that connects ahead of the parallactiscope as shown in Fig. 15.1. This chapter discusses the implementation of rotation modules to permit scene rotation about one, two, or all three deflection axes.

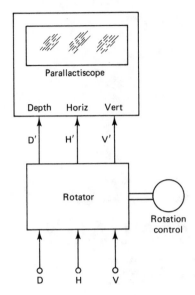

Figure 15.1 Connecting a rotation module to the parallactiscope.

In the case of the parallactiscope, the most useful single-axis rotator would use the H deflection axis as rotation axis. This is because the V axis can be rotated in effect by simply moving to one side, whereas rotation about the D axis does not bring any new parts of the scene into view. The discussion of the single-axis rotator appearing in the next section is written for rotation about the horizontal (H) axis. This means that the H signal, itself, is untransformed. Only the V and D signals are changed as the rotation angle θ varies.

15.2 ROTATING THE 3-D IMAGES

The rotation transformation of Eq. (4.5b), for example, is commonly produced by use of digital techniques. It can also be produced by use of analog techniques. Such techniques are probably more appropriate for use with the parallactiscope and other 3-D oscilloscopes.

Equation (4.1) expanded (multiplied out) using the LT from Eq. (4.5b) gives

$$D' = D \cos \theta - V \sin \theta \qquad \textbf{(15.1a)}$$

$$H' = H \tag{15.1b}$$

$$V' = D \sin \theta + V \cos \theta \tag{15.1c}$$

The block diagram of this system of equations is shown in Fig. 15.2. It appears that four analog multipliers, two trigonometric converters, and two summing devices ("summers") are required to perform rotation about a single axis. The θ-signal is obtained in this case from a linear potentiometer. If a sine-cosine potentiometer is used instead, the required circuitry can be substantially reduced in complexity.

Rotation about a single axis can be produced by use of a 2-gang (dual-cup) sine-cosine potentiometer and a combination of operational amplifiers acting as summers. A simplified schematic of such a rotator is shown in Fig. 15.3. It uses ten operational amplifiers and no multipliers.

Still further simplification of circuitry can be obtained by use of an opto-electronic sine-cosine potentiometer. Such a

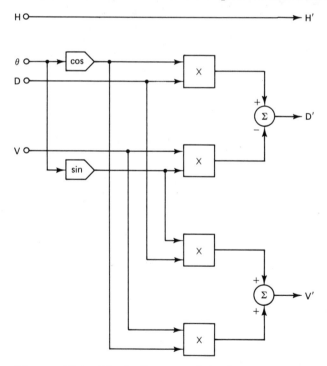

Figure 15.2 Block diagram of single-axis rotator.

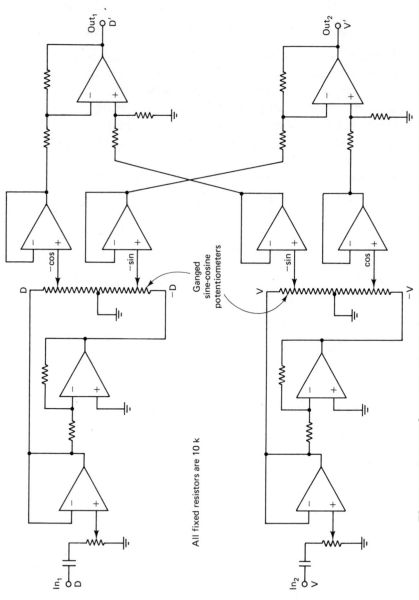

All fixed resistors are 10 k

Figure 15.3 Schematic of single-axis rotator using sine-cosine potentiometers and operational amplifiers.

potentiometer can be constructed from scraps of linear polarizer, a miniature incandescent lamp, and cadmium-sulfide photoconductive cells. A description of this potentiometer and a single-axis rotator using it appears in a later section.

Finally, discrete rotations, as opposed to continuous ones, can be obtained by using a rotary switch to control discrete values of resistance in a rotation circuit. This eliminates the need for potentiometers altogether.

In all these single-axis rotators, the shaft of the rotator control (potentiometer or rotary switch) should be physically oriented to correspond with the image's rotation axis. Thus, if the axis of rotation is H, the potentiometer shaft should be horizontal.

15.3 ROTATION ABOUT MORE THAN ONE AXIS

Rotation about more than one axis is obtained by cascading two or three single-axis rotators. This is shown in the block diagrams of Fig. 15.4. The block in part (a) of the figure represents a rotator of the type discussed in the previous section.[1] Each of the blocks in parts (b) and (c) is of this same type.

Electronically, cascading rotators in this way is conceptually simple. On the other hand, the mechanical coupling of potentiometer shafts is somewhat involved. In Fig. 15.4(b), the second rotator is acting on the primed (') vector, not on the unprimed raw signals as is the first rotator. This means that in order to have the orientations of the rotation axes correspond to those of the image, the second rotator potentiometer must, in effect, be "carried on" the shaft of the first rotator potentiometer. For linear potentiometers (ala Fig. 15.2) and sine-cosine potentiometers (ala Fig. 15.3), this can be accomplished by the mechanism shown in Fig. 15.5 for 2-axis rotation, and by that shown in Fig. 15.6 for 3-axis rotation. These mechanisms consist of combinations of right-angle drives and rotary differentials. For the opto-electronic potentiometer, only right-angle drives are required.

---

[1]Except that its axis is V.

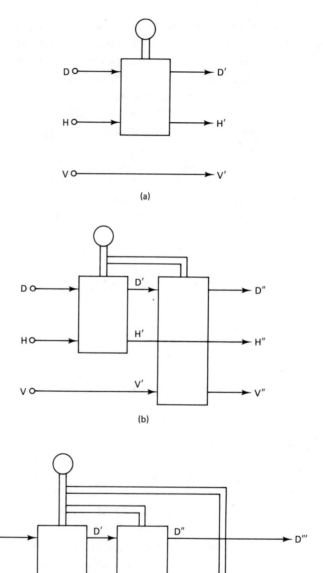

Figure 15.4 Block diagram of rotation: (a) about one axis; (b) about two axes; (c) about three axes.

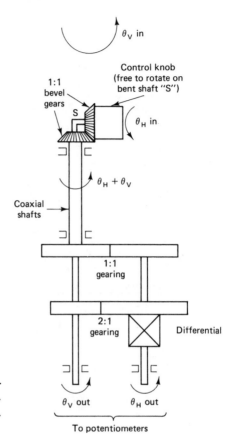

Figure 15.5 How to control two-axis rotation potentiometers with a single coordinated control.

15.4 DESIGNING AN OPTO-ELECTRONIC ROTATOR

In this section, we investigate the possibility of constructing an opto-electronic rotator from linear polarizers to obtain the sine and cosine functions.

The transmittance T for non-polarized light of a pair of identical linear polarizers is

$$T = A \cos^2\theta + B \tag{15.2}$$

where θ is the angle between the two polarizer axes; A and B are constants characteristic of the particular polarizer used. Since $\cos^2\theta = \frac{1}{2} + \frac{1}{2}\cos 2\theta$, Eq. (15.2) can also be written

$$T = A' \cos 2\theta + B' \tag{15.3}$$

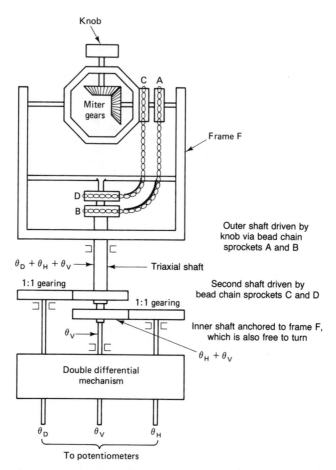

Figure 15.6   Three-axis rotation control.

where A' and B' are two new constants. (Similar expressions apply to the sine function.) This is the desired action. The appearance of "2θ" means that two cycles of transmittance variation occur for each revolution of one polarizer relative to the other.

A primitive opto-electronic potentiometer to obtain the sine function or cosine function is diagrammed in Fig. 15.7. The 2:1 step-down gearing is required because 2θ (instead of θ) appears in Eq. (15.3).

The photocell depicted in Fig. 15.7 must have its resistance linearly dependent on light intensity. What happens in the "real

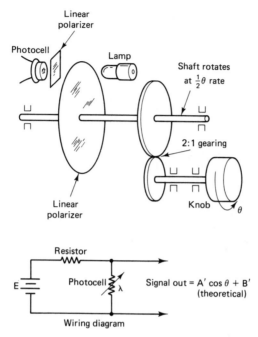

Figure 15.7   A primitive opto-electronic sine or cosine potentiometer.

world," however, is something else. Photoconductive cells that are called "linear" have a characteristic that is linear in terms of conductance. This represents a hyperbolic resistance characteristic; that is, the resistance of such a device is inversely proportional to light intensity. How this is handled is explained next.

The transfer characteristic of the op-amp circuit of Fig. 15.8 is

$$\frac{\text{output}}{\text{input}} = -R_f/R_i \tag{15.4}$$

Thus, if we put the photoconductive cell of Fig. 15.7 in the R_i position, its hyperbolic characteristic produces the desired linear effect on signal output.

The circuit of a practical opto-electronic sine (or cosine) potentiometer is shown in Fig. 15.9. Two photocells are used with one as reference. This helps compensate for lamp aging effects. The author has found that Clairex CL704 photocells with

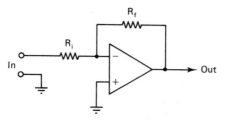

Figure 15.8 An operational
amplifier circuit.

Polaroid HN42 polarizer produces good results. The CL704 cells
have a good linear characteristic in conductance. The HN42
polarizer has a good cyclic light-variation characteristic because
light extinction (for crossed polarizers) is not complete.

The circuit of Fig. 15.9 is one-fourth of a single-axis rotator.
It produces the operation

$$x' = x \sin \theta \tag{15.5}$$

or

$$x' = x \cos \theta \tag{15.6}$$

depending on the orientation of the fixed polarizer attached to
the photocell. By spacing three more pairs of photocells (total of
four pairs) around the periphery of the rotating polarizer, all four
required signals of that type can be obtained. They can then be
combined linearly by two summers to implement Eq. (15.1).

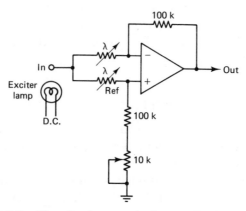

Figure 15.9 Circuit of a practical opto-electronic sine or
cosine potentiometer.

15.5 PERFECTING THE OPTO-ELECTRONIC ROTATOR

Two or three of the single-axis opto-electronic rotators just described can be cascaded to provide manually-controlled image rotation about any two or all three axes. The controls can be ganged as shown in Fig. 15.10 to cause the 3-D images to rotate in the same way as the control is twisted. The control for three axes of rotation is depicted in Fig. 15.11.

In all of these controls for the opto-electronic potentiometer, 2:1 gearing must be used because of Eq. (15.3). This gearing can be eliminated by using a special technique to cause the transfer characteristic of the rotating polarizer to be this instead of Eq. (15.3):

$$T = a \cos \theta + b \qquad (15.7)$$

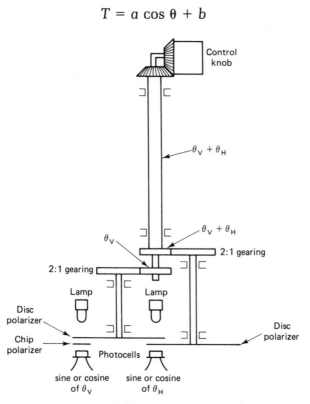

Figure 15.10 Control for two-axis opto-electronic rotator. The two overlapping polarizers perform a differential function.

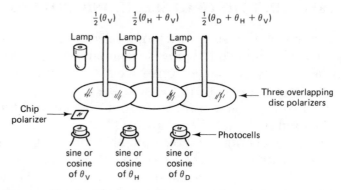

Figure 15.11 Obtaining sine or cosine of θ_V, θ_H, and θ_D from rotations for $\frac{1}{2}\theta_V$, $\frac{1}{2}(\theta_H+\theta_V)$; and $\frac{1}{2}(\theta_D+\theta_H+\theta_V)$.

This end can be accomplished by use of a cone-shaped rotating polarizer instead of the disc-shaped one depicted in Fig. 15.7.

To make such a cone-shaped polarizer, cut a semicircular piece of linear polarizer as shown in Fig. 15.12. Then form it into a cone by butting and cementing the two edges together, producing the cone illustrated in Fig. 15.13. Patient consideration of

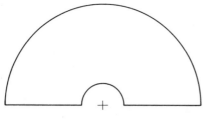

Figure 15.12 Pattern for a cone-shaped polarizer to produce the characteristic $I = a\cos\theta + b$.

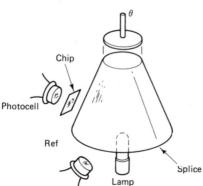

Figure 15.13 Appearance of the cone-shaped polarizer.

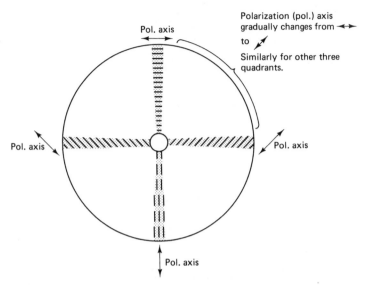

Figure 15.14   A disc-shaped polarizer for producing the characteristic $I = a \cos\theta + b$.

this shape will convince you that the photocell depicted will receive light variation in the manner of Eq. (15.7) instead of Eq. (15.3). Since perfect joining of the two edges to form the cone cannot be obtained, there will be an error signal generated when the splice crosses the photocell position. If you cannot live with this, then rotation can be restricted to less than continuous.

Cone-shaped polarizers can be combined in a manner similar to that for the disc-shaped ones for two- and three-axis rotators.

It may be possible to fabricate disc-shaped linear polarizers having the same rotation characteristic as cone-shaped ones. Such chips would have the polarization pattern indicated schematically in Fig. 15.14. They would permit opto-electronic rotators to be made that are usable continuously through 2 piradians (360°) with no error signals due to splices, and no 2:1 reduction gearing. This would, perhaps, represent the simplest manually-controlled continuous analog rotator that we might reasonably expect to design.

CHAPTER 16

BLUE-SKY PROJECTS FOR YOU TO DO

16.1 A NEXT STEP: ALL-ELECTRONIC PARALLACTISCOPES FOR AIDING SPACE-SHUTTLE RENDEZVOUS AND DOCKING MANEUVERS

Applications for the parallactiscope in its present state of development are probably limited to the laboratory. This is primarily because of the mechanically scanned Collender filter.

If slit scanning can be performed electronically by a solid-state spatial filter, then an all-electronic parallactiscope could be built. Such a parallactiscope could be used outside the laboratory: in classrooms, in commercial and military environments, and in moving vehicles including spacecraft.

Such a solid-state spatial filter was recently reported to have been fashioned and fitted to a parallactiscope by Lowell Noble. It is reportedly small, the images dim, and the range of movement parallax restricted. However, such an accomplishment represents a significant step forward in the development of all electronic real-time holoform displays.

With the availability of a practical all electronic parallactiscope, its application to spacecraft rendezvous and docking displays would become possible. This would have the obvious advantage over present 2-D CRT displays of the added visual dimension; thus aiding in the astronauts' judgment of depth.

16.2 RASTER DISPLAYS FOR TV AND MOVIES: HOW TO PROCEED

Even with the replacement of the mechanically scanned slit with an electronic one, visual image brightness of the parallactiscope would be expected to be less than that of conventional CRT displays. The reason is that much of the light output from the CRT is not used, being blocked by the Collender filter. A way around this limitation can be provided by the use of raster displays instead of the direct-writing oscillographic variety.

In a limiting situation, the raster would consist of a single vertical line that moves to follow the slit exactly. In this way an observer would receive as much light as he would if the slit were not present. While such a situation would be just as bright as a conventional CRT display, there would be no parallax, because as soon as the observer would move to one side, the light would no longer reach him.

A realistic compromise is to use a narrow vertical raster that follows the slit as it scans. This would provide a brighter image than current parallactiscope displays. It would not accommodate as much movement parallax as that provided by current parallactiscope displays; but this total of 1/2 piradian (90°) of movement parallax is really more than would be usable in, say, the cockpit of a space shuttle.

In order to generate shaded 3-D images, such as those required for movies, a raster is required. The form of the raster should be as described earlier in order to maximize visual image brightness.

For TV and movies, in addition to a raster you will need two video-frequency signals: one to provide intensity modulation as with 2-D TV, and one to provide instant-to-instant depth information.

Camera design presents a formidable challenge. It is suggested that you first try a flying-spot camera design, perhaps beginning by modifying an existing flying-spot scanner to provide a depth-video signal in addition to the intensity-video signal. For this a depth sensor would need to be devised. This might take the form of two position sensors coupled in a triangulation mode. From this, a depth signal might be obtained.

The raster generator would need to be redesigned to provide a narrow vertical raster tracking the slit scan on the holoform TV monitor.

Don't be concerned at this time with the seeming "incompatibility" of this scan format with existing TV. Scan formats can be converted to make an incompatible system be compatible, and this will become even more practical as digital TV systems become commonplace.

16.3 COMPUTER-GENERATED *REAL* 3-D WITH THE PARALLACTISCOPE: A SURE BET

Present computer-generated graphics displays use digitally generated signals applied to CRT monitors. The CRT monitors are essentially XY oscilloscopes—usually with an intensity-modulation capability.

The parallactiscope can also be used as an XY oscilloscope. Therefore, those same computer-generated displays can, in principle, be produced using the parallactiscope as the imaging device. But the parallactiscope will accommodate a third deflection input orthogonal to the other two. Therefore, it will accept a third digitally generated signal. In this way real 3-D computer-generated images become possible with the parallactiscope.

If you decide to develop this branch of 3-D graphics, you could conceivably produce 3-D video games, 3-D art, 3-D animated movies, 3-D CAD,[1] and a wide variety of other 3-D computer-generated displays.

It seems certain that such displays will become a reality someday. The parallactiscope is here now, and so the development of such displays can begin immediately.

16.4 ADDING COLOR AND HIGH-DEFINITION 3-D

Use of a multicolor CRT with your parallactiscope will allow you to generate multicolored holoform images. Such an application presents no special difficulties over and above those presented by 2-D color displays and monochrome parallactiscope displays.

High-definition 3-D TV is a logical next step after you have developed holoform TV. You will need to use a smaller CRT spot, a denser raster, and higher-frequency video signals. The problems here are essentially those encountered with "conventional" high-definition TV, except that now you have a second video-frequency signal to be concerned with; namely, the depth video.

16.5 A VARIETY OF THINGS TO INVENT: FLAT-PANEL 3-D, LARGE SCREEN AND *REALLY* LARGE SCREEN 3-D, DOUBLE PARALLACTIC 3-D, AND SO ON

The primary problems you will encounter with the applications listed in the title of this section are as follows.

With flat-panel 3-D, you will need to find an imaging medium with a very short persistence.

With large screen 3-D, you will need to devise a practical large area Collender filter. With "really" large screen 3-D, you may want to consider some kind of projection arrangement.

---

[1]Computer-aided design.

Robert Collender has performed a considerable number of studies on large screen holoform TV and movies.

With double parallax, you will need to consider using a scanning pinhole or crossed slits instead of a single slit. In this way you can get movement parallax both vertically and horizontally. But consider this: If the presence of a slit reduces image intensity, the presence of a pinhole will reduce it even more. You might want to consider obtaining vertical movement parallax by tracking the (single) observer vertically instead of using a pinhole. This can produce the same effect without reducing the image intensity. Even simpler, and perhaps more useful, would be a manually-operated rotation control to permit images to be rotated about the horizontal axis as described in Chapter 15.

The "and so on" in the heading of this section indicates that the list given is not exhaustive. You are invited to devise other "blue-sky projects" of your own!

16.6 4-D FOR DESSERT?

In Chapter 1 (Section 1.6) it was pointed out that certain 2-D Lissajous figures present a 3-dimensional appearance—particularly while drifting. This suggests that drifting 3-D Lissajous figures might present a 4-dimensional appearance—as if they are rigid 4-D figures twisting through some fourth dimension.

It is easy to write the mathematical descriptions of 4-D spaceforms: You simply use four parametric equations. But how would such images be presented for viewing and could anyone really "see" them?

From *Waveforms: A Modern Guide to Nonsinusoidal Waves and Nonlinear Processes:*

> If the display medium is three dimensional . . . four-dimensional spaceforms must be displayed in sections or projections. . . .
>
> Whether or not we can learn to "see" spaceforms of dimensionality greater than three is still open to debate. Since birth, we have been exposed to the three-dimensional objects that surround us. Thus we have learned to see them, that is, to perceive their forms. If we were constantly exposed to

four-dimensional visual objects, could we learn to perceive their forms? Helmholtz is supposed to have believed that such a "hyperperceptive" ability could be developed if the brain is provided the proper inputs. (Henry Parker) Manning reported that he could "almost" see objects of hyperspace as a result of the study of four-dimensional geometry. (A. M.) Noll reported on experiments involving the visualization of N-dimensional spaceforms by use of a digital computer.[2]

Apparently there is no fundamental reason that representations of 4-dimensional spaceforms cannot be displayed so as to convey all the information necessary for hyperspaceform perception.

Present computer-generated 3-D solids graphics are indeed impressive; but they are monocular, and so are not well-suited to the display of 4-dimensional spaceforms. Perhaps real-time holoform displays will one day prove to be a necessary link in the development of a hyperperceptive ability in the human observer.

[2]Ref. 7 in Appendix H.

APPENDIX A

VIEWING THE STEREO PAIRS APPEARING IN THIS BOOK (Ref. Sec. 1.6, 11.3)

If you have a professional stereo viewer available that is compatible with the photo separations in this book, then you can use that to view the stereoscopic and multiscopic views in pairs. They can also be viewed *sans viewing aids* if you have developed that ability. If not, you can easily make the simple lensless viewer shown in Fig. A1.

As shown in Fig. A1, cut the viewer to the indicated dimensions from a manilla folder or similar card stock. The window spacing (dimension marked 63 mm) will accommodate the average observer. This dimension may need to be slightly

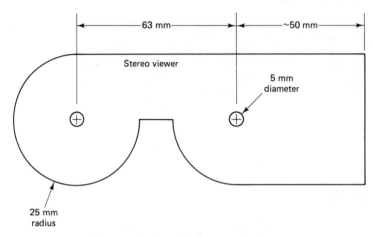

Figure A1 Lensless stereo viewer.

more or less depending on your eye separation. It is suggested that you cut a viewer as shown, determine whether the 63 mm dimension should be increased or decreased, then cut a second viewer using the new window spacing.

To use the viewer, hold it as close to your eyes as possible without bending it. Peer at the stereo pair through the two windows, adjusting your distance from the page for optimum results. The idea is to position yourself (with the viewer) so your left eye sees only one image and your right eye sees only the image on the right. Use your normal reading distance. Relax your vision. It may take a few seconds for you to obtain fusion and experience the depth sensation.

When viewing the displays you can produce on your parallactiscope, no viewing device is required; stereo occurs automatically and naturally, along with movement parallax.

APPENDIX B

PRINTED-CIRCUIT BOARD CONSTRUCTION (Ref. Sec. 9.4)

There is more than one way to prepare a printed-circuit (PC) board. If you have a professional capability, or have such a capability available to you, then the construction of E3 for the parallax adapter (Chapter 9) will present no problem. Otherwise, the following hints may be of help to you. These hints are based on the use of the PC kit listed in Table 9.2 of Chapter 9.

• Preparation of pattern.

    1. Lay out hole pattern full size on tracing paper, placing a cross at each hole center. Also mark the board outline.

2. Cut pattern along the board outline.

● Preparation of copper-clad board.

1. Cut board to size.

2. Overlay pattern onto copper side of board. Fasten it with tape.

3. Gently center punch all hole centers, taking care that the far side of the board does not crack. Remove the pattern.

4. Drill holes, again taking care that the far side of the board does not crack.

5. Clean the copper.

6. Define all pads with resist pen, using the drilled holes as a guide. A drafting circle template does a neat job. Allow the ink to dry after each pad is drawn and filled in.

7. Define conductors and edges of large areas with a resist pen.

8. Cover large areas with plastic electrical tape.

● Etch, following PC kit manufacturer's instructions. Wash.

● Final board preparation.

1. Remove tape.

2. Remove resist and thoroughly clean the board.

3. Insert terminals into holes from the copper side of board as required.

4. The board is now ready to receive angle brackets and electronic components.

NOTE: The writer has found that when making PC boards using a resist pen, reversal of the normal order of etching and drilling generally produces a better product. This is due to the fact that registration between pad and hole is generally better when drilling is done before pads are defined.

APPENDIX C

ADDING THE SLIT TO THE SCANNING PENDULUM (Ref. Sec. 9.5, 11.1)

Cut a sliver 0.10-inch wide by 3-1/2-inches long from a half-wave retarder material (0.030-inch thick plastic material). Cut at 1/4 piradian (45°) to the axis. The edges must be accurately straight and parallel. Alternatively, cut a similar sliver from clear acetate stock. This must be flat; so if you use a 2-liter soft-drink bottle, you must remove its curvature.

In this latter case, cut a 3 × 3-inch square of the material and clamp it between two flat and smooth sheet metal plates. Immerse in boiling water for two to three minutes to flatten it permanently. Expect it to shrink considerably (in addition to becoming flat)!

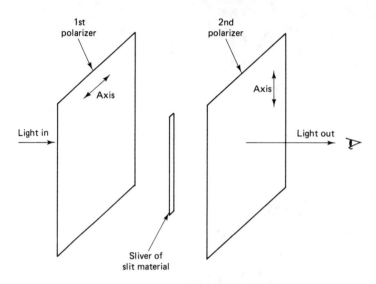

Figure C1 Placement of slit for testing its action.

When you cut the sliver of slit material, be sure the cut is oriented properly. The sliver should let maximum light come through crossed linear polarizers when placed as shown in Fig. C1. Check this orientation before cutting. Check it again afterwards.

Attach the sliver of slit material to the two tines of the slit-scanning pendulum by wrapping with thread and gluing with a small amount of epoxy as shown in Fig. C2. Keep the mass here very low to maximize the scanning frequency. Allow epoxy to cure thoroughly.

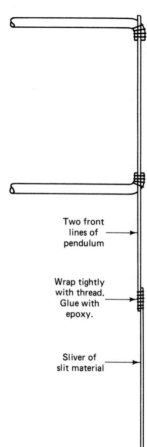

Two front
lines of ————▶
pendulum

Wrap tightly
with thread.
Glue with ————▶
epoxy.

Sliver of ————▶
slit material

Figure C2 How to attach
sliver of slit material to
pendulum.

APPENDIX D

A STATIC-SAFE WORK STATION (Ref. Sec. 10.3)

A static-safe work station has the following equipment and requirements:

1. A grounded, moderately-conductive work surface. This should be connected to a good earth ground via a 1-MΩ resistor. The purpose of the resistor is to limit the current to non-lethal levels in the event that you touch a hot electrical circuit accidentally.
2. A grounded operator. Connect a conductive wrist strap to ground through a 1-MΩ resistor.

3. A grounded-tip soldering iron.

4. Elimination of static from other sources:

   a) All unnecessary nonconductors should be removed from the area.

   b) The operator should wear a short sleeve shirt; no nylon or rayon clothing please!

   c) Other persons should be kept away. Otherwise a grounded conductive floor mat should be provided. Ground via a 1-MΩ resistor.

   d) Air conductivity should be increased by use of an air ionizer or humidifier. Fifty percent humidity is about right.

APPENDIX E

PHOTOGRAPHING PARALLACTISCOPE DISPLAYS (Ref. Sec. 11.3, 13.1, 14.1)

The multiscopic display photos appearing in this book were taken using the camera-positioning fixture shown in Fig. E1.

The fixture consists of two parts pivoted together at the position of the CRT screen. These two parts are shown in Fig. E2. They are a baseplate, which is fixed relative to the parallactiscope, and a camera holder mounted on an arm. The arm is free to swing through angles of 1/4 piradian (45°) both directions from head-on. The position of the arm is marked every 0.05 piradian (9°), counting both directions from the head-on position.

Figure E1   Camera positioning fixture, with parallactiscope and camera attached.

The camera used is a Polaroid 103 with closeup lens. The distance from the front of the lens to the CRT screen is 11.0 inches. The camera holder positions the camera above the arm so its optical axis aligns with the center of the CRT screen.

The camera is loaded with black-and-white film (type 107C), but the controls are set for color. This has the effect of opening the iris to admit more light. The camera is placed on its holder, focused on the CRT screen,[1] and positioned at the desired aspect angle by swinging the arm as required. The room must be totally dark.

Set the parallactiscope controls for a steady image with an intensity near maximum. Time exposures of approximately five

---

[1]This is accomplished by extending the bellows fully for the Polaroid 103.

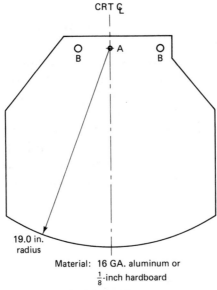

CRT $\mathcal{C}_L$

19.0 in.
radius

Material: 16 GA. aluminum or
$\frac{1}{8}$-inch hardboard

Hole "A": drill for 6-32 machine screw
Hole "B": drill to fit scope feet. Position so hole "A"
is directly under center of CRT screen

(a)

Figure E2a.   Parts for camera holding fixture. Baseplate.

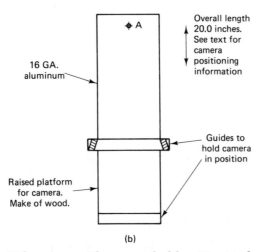

Overall length
20.0 inches.
See text for
camera
positioning
information

16 GA.
aluminum

Guides to
hold camera
in position

Raised platform
for camera.
Make of wood.

(b)

Figure E2b.   Arm with camera holder. Pivot to baseplate
with 6-32 machine screw through hole A.

seconds were used for the pictures in this book. You will want to experiment to determine the optimum time for your particular arrangement.

To obtain time exposures with the Polaroid 103, cover the photocell with black tape. Cock the shutter with the white button as usual. Control the exposure by the length of time you hold the red button down.[2]

The same fixture was used to take the "conventional" scope photos shown in Figs. 1.2, 4.4, 4.5, 4.6, and F4 (in Appendix F). The parallax adapter was removed, the camera placed in the head-on position, and a reduced exposure time (approximately one second) was used for these shots.

[2]If your Polaroid 103 won't take time exposures, the 3-V battery may need to be replaced. To gain access to it, pull open the small door at the left rear of the camera.

APPENDIX F

A SEVEN-HERTZ FREQUENCY METER (Ref. Sec. 11.4)

The scanning pendulum moves a bit too rapidly for the average person to count cycles per second (Hz) by watching or listening to it. If its frequency, f, is divided by two, then a direct visual or aural count becomes possible.

Figure F1 shows a simple instrument that accepts the s-signal from your parallactiscope, divides its frequency by two, and produces an endless train of audible ticks at that $f/2$ rate. Its schematic is shown in Fig. F2.

The circuit uses a 2N2646 unijunction transistor as a pulse generator. The s-signal is used to synchronize it to the scanner

Figure F1   The seven-hertz frequency meter.

oscillation. A 50k linear pot serves to adjust the time constant to the proper range to permit synchronization at the $f/2$ rate. A single power source from 9 to 15 V is required to power the frequency meter. Fifteen volts provides louder ticks.

The proper setting for the 50k pot can be found as follows. Connect the s-signal and have the scanner pendulum operating. Set the 50k pot near its full CCW position and listen to the pulse train. It should be syncopated in such a way that it inserts an extra pulse occasionally (time constant too short). Next set it near its full CW position. The pulses should now be syncopated in such a way that a pulse is lost occasionally (time constant too long). Now find a control setting between those two that produces no syncopation of either kind. This is the "proper" setting. It should exist over approximately one-third of the total control range.

Alternatively, the 50k pot can be set visually by applying outputs to the horizontal and vertical inputs of an XY oscillo-

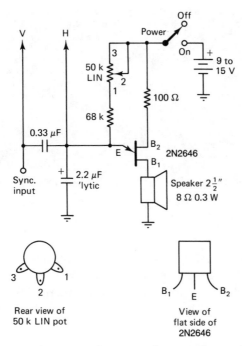

Figure F2 Schematic diagram of seven-hertz frequency meter.

scope as shown in Fig. F3, and adjusting to obtain a "stationary" repetitive pattern of the type shown in Fig. F4.

Once the 50k pot has been set for your parallactiscope, the setting should be marked for future reference.

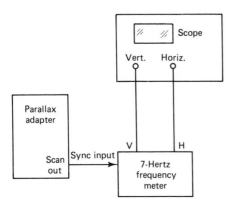

Figure F3 Scope connections to visually set 50k pot.

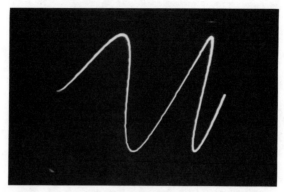

Figure F4 XY oscilloscope display of frequency meter signal using the hookup of Figure F3.

Using the frequency meter is simply a matter of counting audible ticks for 30 seconds or more. You can use Table F1 to find the scanner frequency (f), and the number of slit scans per second ($2f$) when you know the number of ticks in 30 seconds.

TABLE F1. CALIBRATION CHART FOR THE 7-HZ FREQUENCY METER

Number of Ticks in 30 Seconds	Scanner Frequency (f)	Number of Scans per Second ($2f$)
95	6.3 Hz	12.7
96	6.4	12.8
97	6.5	12.9
98	6.5	13.1
99	6.6	13.2
100	6.7	13.3
101	6.7	13.5
102	6.8	13.6
103	6.9	13.7
104	6.9	13.9
105	7.0	14.0
N	N/15	2(N/15)

Alternatively, you can time 100 ticks with a stopwatch and calculate f and $2f$ by use of the following formulas:

$$f = 200/T \text{ hertz}$$
$$2f = 400/T \text{ scans per second}$$

where T is the number of seconds for 100 ticks.

The circuit of Fig. F2 has a free-running frequency range of about 85 to 150 pulses per thirty seconds.

APPENDIX G

SUPPLIERS' NAMES AND ADDRESSES (Ref. Sec. 7.4, 9.2, 10.2)

Electronic components

Albia Electronics
44 Kendall Street
New Haven, CT 06508

Burr-Brown Research Corp.
P.O. Box 11400
Tucson, AZ 85734

Hanifin Electronics Corp.
P.O. Box 188
Bridgeport, PA 19405

Keystone Electronics Corp.
49 Bleecker Street
New York, NY 10012

Mouser Electronics
11433 Woodside Avenue
Santee, CA 92071

Polarizing materials

Edmund Scientific
101 E. Gloucester Pike
Barrington, NJ 08007

Meadowlark Optics
7460 East County Line Road
Longmont, CO 80501

Polaroid Corporation
Cambridge, MA 02139

APPENDIX H

BIBLIOGRAPHY

CRTS AND OSCILLOSCOPES

1. *The Oscillographer*, Vol. 12, No. 4, Oct-Dec, 1951, Allen B. Du Mont Laboratories, Inc., Instrument Division, 750 Bloomfield Ave., Clifton, NJ 07015. This issue contained an announcement of "some typical special tubes." Among these was one containing four wholly independent electron-gun and deflection-electrode structures, and one usable at frequencies up to 1000 MHz. This fine periodical publication is, unfortunately, out of print.

2. Lenk, J. D., *Handbook of Oscilloscopes: Theory and Application*, Prentice-Hall, Inc., Englewood Cliffs, NJ, 1968.

3. Rider, J. F., *The Cathode-Ray Tube at Work*, John F. Rider Publisher, Inc., 1938, 21st printing 1945. A classic. Rider Publisher was sold to Hayden Book Co. Pioneer author Rider died Feb. 6, 1985 at the age of 85.

4. Smith, P. C., *Know Your Oscilloscope*, Howard W. Sams & Co., Inc., The Bobbs-Merrill Co., Inc., NY, Second Edition, 1967.

5. Soller, T., M. A. Starr, and G. E. Valley, Jr. (eds), *Cathode Ray Tube Displays*, No. 22 in the Massachusetts Institute of Technology Radiation Laboratory Series. Reprint by Boston Technical Publishers, Inc., 5 Bryant Road, Lexington, MA 02173, 1964. This fine book chronicles the work done at MIT during World War II.

DESIGN

6. Johnson, D. E. and J. L. Hilburn, *Rapid Practical Design of Active Filters*. J. Wiley & Sons, NY, 1975.

7. Tilton, H. B., *Waveforms: A Modern Guide to Nonsinusoidal Waves and Nonlinear Processes*. Prentice-Hall, Englewood Cliffs, NJ, 1986.

THE PULFRICH STEREOPHENOMENON

8. MacDonald, R. I., "Three-dimensional television by texture parallax," *Applied Optics*, Vol. 17, No. 2, 15 January 1978, pp. 168–70.

    An application of the Pulfrich stereophenomenon to obtain stereo depth from single pictures.

9. Tilton, H. B., "Pulfrich Space Form," 7th National Symposium on Information Display *Technical Session Proceedings*, Society for Information Display, 1966, pp. 223–32.

    A quantitative discussion of the Pulfrich effect.

SPATIAL IMAGING

10. Abramson, A., "Stereoscopic Color Television System," U.S. Patent 2,931,855, April 5, 1960.
   Camera method and CRT using lenticular screen.

11. "Sculpting via Computer Optics," *American Machinist*. Sept. 1979, pp. 79–82.

12. "Flight Patterns Displayed in 3-D," *Aviation Week*. May 30, 1960, p. 85.

13. Bedford, A. V., "Television System," U.S. Patent 2,307,188, Jan. 5, 1943.

14. Berkley, C., "Three-Dimensional Representations on Cathode-Ray Tubes," *Proc. IRE*, Vol. 36, No. 12, Dec. 1948, pp. 1530–35.

15. Butterfield, J. F., "Stereo Television Microscope," U.S. Patent 3,818,125, June 18, 1974.

16. Carnahan, C. W., "Stereoscopic Method and Apparatus," U.S. Patent 2,301,254, Nov. 10, 1942.

17. Collender, R. B., "Three Dimensional Unaided Viewing Method and Apparatus," U.S. Patent 3,178,720, April 13, 1965.

18. Collender, R. B., "The Stereoptiplexer," *Information Display*. November/December 1967, pp. 27–31.

19. Collender, R. B., "Standard Theatre Stereoptics Without Glasses," *Information Display*. July/August 1972, pp. 18–25.

20. Collier, R. J., "Holography and Integral Photography," *Physics Today*. Vol. 21, No. 7, July 1968, p. 54.

21. De Rosa, L. A., "Method of and Means for Visually Reproducing Signals," U.S. Patent 2,408,050, Sept. 24, 1946.
   A stereoscopic television system.

22. "Solids Engine Enhanced for Imaging," *Electronic Imaging*. March 1985, pp. 12, 13.

23. "Du Mont Forecasts TV Transmissions to the Blind," *Electronic Industries*. Vol. 22, No. 7, July 1963, p. 150.

24. "Sylvania Shows 3-D TV System," *Electronic News*. March 24, 1964.

25. "3-D Scope," *Electronic Products*. Dec. 1965, pp. 24–25.

26. "CRT Provides Three-Dimensional Displays," *Electronics.* Nov. 2, 1962, pp. 54–57.

27. "Engineering Highlights: 1962 IRE Convention," *Electronics.* March 9, 1962, pp. 51*ff.*

28. "Three-Dimensional CRT Uses Atomic Resonance," *Electronics.* Jan. 11, 1963, pp. 52–54.

29. "Engine Enables PC-AT to Draw 3-D Images," *Electronics.* Vol. 58, No. 33, August 19, 1985, p. 54.

30. "3-D Makes CAD Lively," *ElectronicsWeek.* Jan. 14, 1985, pp. 49–50.

31. Ferguson, E. T., "Some Notes on Stereoscopic Display, and an Isochromic Anaglyph C.R.T.," *9th National Symposium on Information Display Technical Session Proceedings.* Society for Information Display, 654 North Sepulveda Boulevard, Los Angeles, CA 90049, 1968, pp. 201–7.

32. Françon, M., *Holography,* Academic Press, NY, 1974.

33. Geer, C. W., "Three-dimensional Display Apparatus," U.S. Patent 3,184,630, May 18, 1965.

    CRT using a lenticular screen.

34. Gerritsen, H. J., "Various Design Considerations of Three-Dimensional Photographic as Well as Synthesized Displays Using Fly's Eye Lenslets," *Proceedings of the Electro-Optical Systems Design Conference.* Sept. 1969, pp. 642–49.

35. "Researching Holography," *International Television.* July, 1985, p. 46.

    A discussion of Ronald Kirk's optical tunnel array.

36. Iwane, W., "Three-dimensional Television System," U.S. Patent 4,062,045, Dec. 6, 1977.

    Uses a "plurality" of TV cameras to collect images from several different angles.

37. King, M. C. and D. H. Berry, "Varifocal Mirror Technique for Video Transmission of Three-Dimensional Images," *Applied Optics.* Vol. 9, No. 9, Sept. 1970, pp. 2035–39.

38. Kurtz, R. L., "Real-time Moving Scene Holographic Camera System," U.S. Patent 3,752,556, Aug. 14, 1973.

    A holographic motion picture camera system.

39. Lewis, J. D., C. M. Verber, and R. B. McGhee, "A True Three-Dimensional Display," *IEEE Transactions on Electron Devices.* Vol. ED-18, No. 9, Sept. 1971, pp. 724–32.

40. "Electronic Pathways in Synthetic Worlds," *Machine Design.* Vol. 37, No. 15, June 24, 1965, p. 133.

41. De Montebello, R. L., "Wide-angle Integral Photography— The Integram® System," *Proceedings of the Society of Photo-Optical Instrumentation Engineers.* Volume 120, Three-Dimensional Imaging, 1977, pp. 73–91.

42. "Screen of Cylindrical Lenses Produces Stereoscopic Television Pictures," *NASA Tech Brief #66-10086,* March, 1966.

43. "3-D Recorder," *Electro-Technology.* Sept. 1968, pp. 24, 25.

44. Okashi, T., *Three-Dimensional Imaging Techniques.* Academic Press, NY, 1976, 403pp.

45. Parker, E. and P. R. Wallis, "Three-Dimensional Cathode-Ray Tube Displays," *Journal of the IEE.* London, Vol. 95, Part III, No. 37, Sept. 1948, pp. 371–90.

46. "LCD Goes 3-D," *Photonics Spectra.* June 1985, p. 44.

47. "Stereoscope/Computer Projects 3-D Image," *Product Engineering.* Oct. 9, 1967, p. 174.

48. Rawson, E. G., "3-D Computer-Generated Movies Using a Varifocal Mirror," *Applied Optics.* Vol. 7, No. 8, Aug. 1968, pp. 1505–11.

49. Roese, J. A., "Liquid Crystal Stereoscopic Television System," U.S. Patent 3,821,466, June 28, 1974.
    An underwater stereoscopic viewing system.

50. Savoye, F., "Equipment for the Projection of Stereoscopic Views and Films," U.S. Patent 2,421,393, June 3, 1947.

51. Schmitt, O. H., "Cathode-Ray Presentation of Three-Dimensional Data," *J. of Applied Physics.* Vol. 18, Sept. 1947, pp. 819–29.

52. Tilton, H. B., "3-D Display," *Instruments and Control Systems.* Aug. 1966, pp. 83–85.

53. Tilton, H. B., "Generate Three-Dimensional Stereoscopic Patterns on your Oscilloscope," *Radio-Electronics.* In press.

54. Tilton, H. B., "Principles of 3-D CRT Displays," *Control Engineering.* Feb. 1966, pp. 74–78.

55. Tilton, H. B., "Principles of Scenographic CRT Displays," *Automation in Electronic Test Equipment*. Vol. IV, Edited by D. M. Goodman, New York University Press, April 1967, pp. 347–74.

56. Tilton, H. B., "Real-time Direct-viewed CRT Displays Having Holographic Properties," *Proceedings of the Technical Program, Electro-optical Systems Design Conference—1971 West*. Cahners Exposition Group, 1350 East Touhy Avenue, Des Plaines, IL 60017, 1971.

57. Tilton, H. B., *Instruction Manual for Developmental Parallactiscope Model D8105 Style 1983*. Visonics Laboratories, 8401 Desert Steppes Drive, Tucson, AZ 85710, 1983.

58. Traub, A. C., "Stereoscopic Display Using Rapid Varifocal Mirror Oscillations," *Applied Optics*. Vol. 6, No. 6, June 1967, pp. 1085–87.

59. Tripp, R. M., "Three-Dimensional Television," U.S. Patent 3,932,699, Jan. 13, 1976.

Lenticular screen system.

60. Vanderhooft, J. J., "Stereoscopic Television Means," U.S. Patent 2,783,406, Feb. 26, 1957.

A proposed CRT using a lenticular screen.

61. Whittaker, J. L., "Three-Dimensional Oscillograph System," U.S. Patent 2,455,456, Dec. 7, 1948.

An early stereo oscilloscope and an early monocular 3-D display.

62. Zworykin, V. K., "Television System," U.S. Patent 2,107,464, Feb. 8, 1938. Description of a stereoscopic television system. Filed in 1932.

VISUAL SPACE PERCEPTION

63. Graham, C. H. (ed), *Vision and Visual Perception*. John Wiley & Sons, Inc., NY, 1965.

64. Tilton, H. B., "A Matrix Formulation of Visual Space Perception," *Information Display*. Vol. 4, No. 1, Jan./Feb. 1967, pp. 29–33.

INDEX